ISAAC ASIMOV is undoubtedly America's foremost writer on science for the layman. An Associate Professor of Biochemistry at the Boston University School of Medicine, he has written well over a hundred books, as well as hundreds of articles in publications ranging from *Esquire* to Atomic Energy Commission pamphlets. Famed for his science fiction writing (his three-volume Hugo Award-winning THE FOUNDATION TRILOGY is available in individual Avon editions and as a one-volume Equinox edition), Dr. Asimov is equally acclaimed for such standards of science reportage as THE UNIVERSE, LIFE AND ENERGY, THE SOLAR SYSTEM AND BACK, ASIMOV'S BIOGRAPHICAL ENCYCLOPEDIA OF SCIENCE AND TECHNOLOGY, and ADDING A DIMENSION (all available in Avon editions). His non-science writings include the two-volume ASIMOV'S GUIDE TO SHAKESPEARE, ASIMOV'S ANNOTATED DON JUAN, and the two-volume ASIMOV'S GUIDE TO THE BIBLE (available in a two-volume Avon edition). Born in Russia, Asimov came to this country with his parents at the age of three, and grew up in Brooklyn. In 1948 he received his Ph.D. in Chemistry at Columbia and then joined the faculty at Boston University, where he works today.

Other Avon Books by
Isaac Asimov

ASIMOV'S GUIDE TO THE BIBLE, THE NEW TESTAMENT	24786	$4.95
ASIMOV'S GUIDE TO THE BIBLE, THE OLD TESTAMENT	24794	4.95
ADDING A DIMENSION	22673	1.25
FROM EARTH TO HEAVEN	29231	1.75
THE FOUNDATION TRILOGY	26930	4.95
LIFE AND ENERGY	31666	1.95
THE NEUTRINO	25544	1.50
SOLAR SYSTEM AND BACK	10157	1.25
VIEW FROM A HEIGHT	24547	1.25
FOUNDATION	29579	1.50
FOUNDATION AND EMPIRE	30627	1.50
SECOND FOUNDATION	29280	1.50
THE FOUNDATION TRILOGY BOXED SET	30288	4.50
THE UNIVERSE: FROM FLAT EARTH TO QUASAR	27979	1.95

OF TIME AND SPACE AND OTHER THINGS

ISAAC ASIMOV

A DISCUS BOOK/PUBLISHED BY AVON BOOKS

All essays in this volume are reprinted from
THE MAGAZINE OF FANTASY AND SCIENCE FICTION.

AVON BOOKS
A division of
The Hearst Corporation
959 Eighth Avenue
New York, New York 10019

Copyright © 1965 by Isaac Asimov
Copyright © 1959, 1963, 1964 by Mercury Press, Inc.
Published by arrangement with Doubleday and Company, Inc.
Library of Congress Catalog Card Number: 65-17259
ISBN: 0-380-00325-2

All rights reserved, which includes the right
to reproduce this book or portions thereof in
any form whatsoever. For information address
Doubleday and Company, Inc., 277 Park Avenue,
New York, New York 10017.

First Discus Printing, May, 1975.
Third Printing

DISCUS TRADEMARK REG. U.S. PAT. OFF. AND
FOREIGN COUNTRIES, REGISTERED TRADEMARK—
MARCA REGISTRADA, HECHO EN CHICAGO, U.S.A.

Printed in the U.S.A.

CONTENTS

To Wendy Weil—
with a sigh

INTRODUCTION

As we trace the development of man over the ages, it seems in many respects a tale of glory and victory; of the development of the brain; of the discovery of fire; of the building of cities and civilizations; of the triumph of reason; of the filling of the Earth and of the reaching out to sea and space.

But increasing knowledge leads not to conquest only, but to utter defeat as well, for one learns not only of new potentialities, but also of new limitations. An explorer may discover a new continent, but he may also stumble over the world's end.

And it is so with mankind. We are distinguished from all other living species by our power over the inanimate universe; and we are distinguished from them also by our abject defeat by the inanimate universe, for we alone have learned of defeat.

Consider that no other species (as far as we know) can possess our concept of time. An animal may remember, but surely it can have no notion of "past" and certainly not of "future."

No non-human creature lives in anything but the present moment. No non-human creature can foresee the inevitability of its own death. Only man is mortal, in the sense that only man is aware that he is mortal.

Robert Burns said it better in his poem *To a Mouse*. He addresses the mouse, after turning up its nest with his plough, apologizing to it for the disaster he has brought upon it, and reminding it fatalistically that "The best-laid schemes o' mice and men / Gang aft a-gley."

But then, in a final soul-chilling stanza (too often lost in the glare of the much more famous penultimate stanza about mice and men), he gets to the real nub of the poem and says:

> "Still thou art blest compar'd wi' me!
> "The present only toucheth thee:
> "But oh, I backward cast my e'e
> "On prospects drear!
> "An' forward tho' I canna see,
> "I guess an' fear!"

Somewhere, then, in the progress of evolution from mouse to man, a primitive hominid first caught and grasped at the notion that someday he would die. Every living creature died at last, our proto-philosopher could not help but notice, and the great realization somehow dawned upon him that he himself would do so, too. If death must come to all life, it must come to himself as well, and ahead of him he saw world's end.

We talk often about the discovery of fire, which marked man off from all the rest of creation. Yet the discovery of death is surely just as unique and may have been just as driving a force in man's upward climb.

The details of both discoveries are lost forever in the shrouded and impenetrable fog of pre-history, but they appear in myths. The discovery of fire is celebrated most famously in the Greek myth of Prometheus, who stole fire from the Sun for the poor, shivering race of man.

And the discovery of death is celebrated most famously in the Hebrew myth of the Garden of Eden, where man first dwelt in the immortality that came of the ignorance of time. But man gained knowledge, or, if you prefer, he ate of the fruit of the tree of knowledge of good and evil.

And with knowledge, death entered the world, in the sense that man knew he must die. In biblical terms, this awareness of death is described as resulting from divine revelation. In the solemn speech in which He apprises Adam of the punishment for disobedience, God tells him

(Gen. 3:19): "... for dust thou art, and unto dust shalt thou return."

But man struggles onward under the terrible weight of Adam's curse, and I cannot help but wonder how much of man's accomplishment traces directly back to his endeavor to neutralize the horrifying awareness of inevitable death. He may transfer the consciousness of existence from himself to his family and find immortality after all in the fact that though his own spark of life snuffs out, an allied spark continues in the children that issued out of his body. How much of tribal society is based on this?

Or he may decide that the true life is not of the body which is, indeed, mortal and must suffer death; but of the spirit which lives forever. And how much of philosophy and religion and the highest aspirations of man's faculties arises from this striving to deny Adam's curse?

Yet what of a society in which the notion of family and of spirit weakens; a society in which the material world of the senses gradually fills the consciousness from horizon to horizon? The nearest approach to such a society in man's history is probably our own. How, then, has the modern West, which has deprived itself of the classical escapes, reacted to the inevitability of death?

Is it entirely a coincidence that of all cultures, that of the present-day West is the most time-conscious? That it has spent more of its energies in studying time, measuring time, cutting time up into ever-tinier segments with ever-greater accuracy?

Is it entirely a coincidence that the most materialistic subdivision of our most materialistic culture, the twentieth-century American, is never seen anywhere without his wristwatch? At no time, apparently, dare he be unaware of the sweep of the second hand and of the ticks that mark off the inexorable running out of the sands of his life.

So it is that the opening essays in this collection deal directly with man's attempt to measure time. The notion of time creeps into a number of the other essays as well; in a discussion of units which turn out always to include the "second"; in a discussion of catalysts which squeeze more action into less time. For really, time is a subject that cannot be entirely excluded from any corner of science.

When man faces death directly, then, he studies time, for it is by accurately handling time that he can measure other phenomena and find a route through science. And through science, perhaps, may come a truly materialistic defeat of Adam's curse.

For my final essay in this book takes up the inevitability of death, and the conclusion is that though all men are mortal, they are not nearly as mortal as they ought to be.

Why not? That is the chink in death's armor. Why does man live as surprisingly long as he does? If we can someday find the answer to that, we may find the answer to much more.

Immortality?

Who knows, but—maybe!

Part I
OF TIME AND SPACE

1. THE DAYS OF OUR YEARS

A group of us meet for an occasional evening of talk and nonsense, followed by coffee and doughnuts and one of the group scored a coup by persuading a well-known entertainer to attend the session. The well-known entertainer made one condition, however. He was not to entertain, or even be asked to entertain. This was agreed to.

Now there arose a problem. If the meeting were left to its own devices, someone was sure to begin badgering the entertainer. Consequently, other entertainment had to be supplied, so one of the boys turned to me and said, "Say, you know what?"

I knew what and I objected at once. I said, "How can I stand up there and talk with everyone staring at this other fellow in the audience and wishing *he* were up there instead? You'd be throwing me to the wolves!"

But they all smiled very toothily and told me about the wonderful talks I give. (Somehow everyone quickly discovers the fact that I soften into putty as soon as the flattery is turned on.) In no time at all, I agreed to be thrown to the wolves. Surprisingly, it worked, which speaks highly for the audience's intellect—or perhaps their magnanimity.

As it happened, the meeting was held on "leap day" and so my topic of conversation was ready-made and the gist of it went as follows:

I suppose there's no question but that the earliest unit of time-telling was the day. It forces itself upon the awareness of even the most primitive of humanoids. However, the day is not convenient for long intervals of time. Even allowing a primitive life-span of thirty years, a man would live some 11,000 days and it is very easy to lose track among all those days.

Since the Sun governs the day-unit, it seems natural to turn to the next most prominent heavenly body, the Moon, for another unit. One offers itself at once, ready-made—the period of the phases. The Moon waxes from nothing to a full Moon and back to nothing in a definite period of time. This period of time is called the "month" in English (clearly from the word "moon") or, more specifically, the "lunar month," since we have other months, representing periods of time slightly shorter or slightly longer than the one that is strictly tied to the phases of the moon.

The lunar month is roughly equal to 29½ days. More exactly, it is equal to 29 days, 12 hours, 44 minutes, 2.8 seconds, or 29.5306 days.

In pre-agricultural times, it may well have been that no special significance attached itself to the month, which remained only a convenient device for measuring moderately long periods of time. The life expectancy of primitive man was probably something like 350 months, which is a much more convenient figure than that of 11,000 days.

In fact, there has been speculation that the extended lifetimes of the patriarchs reported in the fifth chapter of the Book of Genesis may have arisen out of a confusion of years with lunar months. For instance, suppose Methuselah had lived 969 lunar months. This would be just about 79 years, a very reasonable figure. However, once that got twisted to 969 years by later tradition, we gained the "old as Methuselah" bit.

However, I mention this only in passing, for this idea is not really taken seriously by any biblical scholars. It is much more likely that these lifetimes are a hangover from Babylonian traditions about the times before the Flood. . . . But I am off the subject.

It is my feeling that the month gained a new and enhanced importance with the introduction of agriculture.

An agricultural society was much more closely and precariously tied to the seasons than a hunting or herding society was. Nomads could wander in search of grain or grass but farmers had to stay where they were and hope for rain. To increase their chances, farmers had to be certain to sow at a proper time to take advantage of seasonal rains and seasonal warmth; and a mistake in the sowing period might easily spell disaster. What's more, the development of agriculture made possible a denser population, and that intensified the scope of the possible disaster.

Man had to pay attention, then, to the cycle of seasons, and while he was still in the prehistoric stage he must have noted that those seasons came full cycle in roughly twelve months. In other words, if crops were planted at a particular time of the year and all went well, then, if twelve months were counted from the first planting and crops were planted again, all would again go well.

Counting the months can be tricky in a primitive society, especially when a miscount can be ruinous, so it isn't surprising that the count was usually left in the hands of a specialized caste, the priesthood. The priests could not only devote their time to accurate counting, but could also use their experience and skill to propitiate the gods. After all, the cycle of the seasons was by no means as rigid and unvarying as was the cycle of day and night or the cycle of the phases of the moon. A late frost or a failure of rain could blast that season's crops, and since such flaws in weather were bound to follow any little mistake in ritual (at least so men often believed), the priestly functions were of importance indeed.

It is not surprising then, that the lunar month grew to have enormous religious significance. There were new Moon festivals and special priestly proclamations of each one of them, so that the lunar month came to be called the "synodic month."

The cycle of seasons is called the "year" and twelve lunar months therefore make up a "lunar year." The use of lunar years in measuring time is referred to as the use of a "lunar calendar." The only important group of people in modern times, using a strict lunar calendar, are the Mohammedans. Each of the Mohammedan years is made up of 12 months which are, in turn, usually made up of 29 and 30 days in alternation.

Such months average 29.5 days, but the length of the true lunar month is, as I've pointed out, 29.5306 days. The lunar year built up out of twelve 29.5-day months is 354 days long, whereas twelve lunar months are actually 354.37 days long.

You may say "So what?" but don't. A true lunar year should always start on the day of the new Moon. If, however, you start one lunar year on the day of the new Moon and then simply alternate 29-day and 30-day months, the third year will start the day before the new Moon, and the sixth year will start two days before the new Moon. To properly religious people, this would be unthinkable.

Now it so happens that 30 true lunar years come out to be almost exactly an even number of days—10,631.016. Thirty years built up out of 29.5-day months come to 10,620 days—just 11 days short of keeping time with the Moon. For that reason, the Mohammedans scatter 11 days through the 30 years in some fixed pattern which prevents any individual year from starting as much as a full day ahead or behind the new Moon. In each 30-year cycle there are nineteen 354-day years and eleven 355-day years, and the calendar remains even with the Moon.

An extra day, inserted in this way to keep the calendar even with the movements of a heavenly body, is called an "intercalary day"; a day inserted "between the calendar," so to speak.

The lunar year, whether it is 354 or 355 days in length, does not, however, match the cycle of the seasons. By the dawn of historic times the Babylonian astronomers had noted that the Sun moved against the background of stars (see Chapter 4). This passage was followed with absorption because it grew apparent that a complete circle of the sky by the Sun matched the complete cycle of the seasons closely. (This apparent influence of the stars on the seasons probably started the Babylonian fad of astrology—which is still with us today.)

The Sun makes its complete cycle about the zodiac in roughly 365 days, so that the lunar year is about 11 days shorter than the season-cycle, or "solar year." Three lunar years fall 33 days, or a little more than a full month behind the season-cycle.

This is important. If you use a lunar calendar and start it so that the first day of the year is planting time, then three years later you are planting a month too soon, and by the time a decade has passed you are planting in midwinter. After 33 years the first day of the year is back where it is supposed to be, having traveled through the entire solar year.

This is exactly what happens in the Mohammedan year. The ninth month of the Mohammedan year is named Ramadan, and it is especially holy because it was the month in which Mohammed began to receive the revelation of the Koran. In Ramadan, therefore, Moslems abstain from food and water during the daylight hours. But each year, Ramadan falls a bit earlier in the cycle of the seasons, and at 33-year intervals it is to be found in the hot season of the year; at this time abstaining from drink is particularly wearing, and Moslem tempers grow particularly short.

The Mohammedan years are numbered from the Hegira; that is, from the date when Mohammed fled from Mecca to Medina. That event took place in A.D. 622. Ordinarily, you might suppose, therefore, that to find the number of the Mohammedan year, one need only subtract 622 from the number of the Christian year. This is not quite so, since the Mohammedan year is shorter than ours. I write this chapter in A.D. 1964 and it is now 1342 solar years since the Hegira. However, it is 1384 lunar years since the Hegira, so that, as I write, the Moslem year is A.H. 1384.

I've calculated that the Mohammedan year will catch up to the Christian year in about nineteen millennia. The year A.D. 20,874 will also be A.H. 20,874, and the Moslems will then be able to switch to our year with a minimum of trouble.

But what can we do about the lunar year in order to make it keep even with the seasons and the solar year? We can't just add 11 days at the end, for then the next year would not start with the new Moon and to the ancient Babylonians, for instance, a new Moon start was essential.

However, if we start a solar year with the new Moon and wait, we will find that the twentieth solar year thereafter starts once again on the day of the new Moon. You see, 19 solar years contain just about 235 lunar months.

Concentrate on those 235 lunar months. That is equivalent to 19 lunar years (made up of 12 lunar months each) plus 7 lunar months left over. We could, then, if we wanted to, let the lunar years progress as the Mohammedans do, until 19 such years had passed. At this time the calendar would be exactly 7 months behind the seasons, and by adding 7 months to the 19th year (a 19th year of 19 months—very neat) we could start a new 19-year cycle, exactly even with both the Moon and the seasons.

The Babylonians were unwilling, however, to let themselves fall 7 months behind the seasons. Instead, they added that 7-month discrepancy through the 19-year cycle, one month at a time and as nearly evenly as possible. Each cycle had twelve 12-month years and seven 13-month years. The "intercalary month" was added in the 3rd, 6th, 8th, 11th, 14th, 17th, and 19th year of each cycle, so that the year was never more than about 20 days behind or ahead of the Sun.

Such a calendar, based on the lunar months, but gimmicked so as to keep up with the Sun, is a "lunar-solar calendar."

The Babylonian lunar-solar calendar was popular in ancient times since it adjusted the seasons while preserving the sanctity of the Moon. The Hebrews and Greeks both adopted this calendar and, in fact, it is still the basis for the Jewish calendar today. The individual dates in the Jewish calendar are allowed to fall slightly behind the Sun until the intercalary month is added, when they suddenly shoot slightly ahead of the Sun. That is why holidays like Passover and Yom Kippur occur on different days of the civil calendar (kept strictly even with the Sun) each year. These holidays occur on the same day of the year each year in the Jewish calendar.

The early Christians continued to use the Jewish calendar for three centuries, and established the day of Easter on that basis. As the centuries passed, matters grew somewhat complicated, for the Romans (who were becoming Christian in swelling numbers) were no longer used to a lunar-solar calendar and were puzzled at the erratic jumping about of Easter. Some formula had to be found by which the correct date for Easter could be calculated in advance, using the Roman calendar.

It was decided at the Council of Nicaea, in A.D. 325 (by which time Rome had become officially Christian), that Easter was to fall on the Sunday after the first full Moon after the vernal equinox, the date of the vernal equinox being established as March 21. However, the full Moon referred to is not the actual full Moon, but a fictitious one called the "Paschal Full Moon" ("Paschal" being derived from *Pesach*, which is the Hebrew word for Passover). The date of the Paschal Full Moon is calculated according to a formula involving Golden Numbers and Dominical Letters, which I won't go into.

The result is that Easter still jumps about the days of the civil year and can fall as early as March 22 and as late as April 25. Many other church holidays are tied to Easter and likewise move about from year to year.

Moreover, all Christians have not always agreed on the exact formula by which the date of Easter was to be calculated. Disagreement on this detail was one of the reasons for the schism between the Catholic Church of the West and the Orthodox Church of the East. In the early Middle Ages there was a strong Celtic Church which had its own formula.

Our own calendar is inherited from Egypt, where seasons were unimportant. The one great event of the year was the Nile flood, and this took place (on the average) every 365 days. From a very early date, certainly as early as 2781 B.C., the Moon was abandoned and a "solar calendar," adapted to a constant-length 365-day year, was adopted.

The solar calendar kept to the tradition of 12 months, however. As the year was of constant length, the months were of constant length, too—30 days each. This meant that the new Moon could fall on any day of the month, but the Egyptians didn't care. (A month not based on the Moon is a "calendar month.")

Of course 12 months of 30 days each add up only to 360 days, so at the end of each 12-month cycle, 5 additional days were added and treated as holidays.

The solar year, however, is not exactly 365 days long. There are several kinds of solar years, differing slightly in length, but the one upon which the seasons depend is the "tropical year," and this is about 365¼ days long.

This means that each year, the Egyptian 365-day year falls ¼ day behind the Sun. As time went on the Nile flood occurred later and later in the year, until finally it had made a complete circuit of the year. In 1460 tropical years, in other words, there would be 1461 Egyptian years.

This period of 1461 Egyptian years was called the "Sothic cycle," from Sothis, the Egyptian name for the star Sirius. If, at the beginning of one Sothic cycle, Sirius rose with the Sun on the first day of the Egyptian year, it would rise later and later during each succeeding year until finally, 1461 Egyptian years later, a new cycle would begin as Sothis rose with the Sun on New Year's Day once more.

The Greeks had learned about that extra quarter day as early as 380 B.C., when Eudoxus of Cnidus made the discovery. In 239 B.C. Ptolemy Euergetes, the Macedonian king of Egypt, tried to adjust the Egyptian calendar to take that quarter day into account, but the ultra-conservative Egyptians would have none of such a radical innovation.

Meanwhile, the Roman Republic had a lunar-solar calendar, one in which an intercalary month was added every once in a while. The priestly officials in charge were elected politicians, however, and were by no means as conscientious as those in the East. The Roman priests added a month or not according to whether they wanted a long year (when the other annually elected officials in power were of their own party) or a short one (when they were not). By 46 B.C., the Roman calendar was 80 days behind the Sun.

Julius Caesar was in power then and decided to put an end to this nonsense. He had just returned from Egypt where he had observed the convenience and simplicity of a solar year, and imported an Egyptian astronomer, Sosigenes, to help him. Together, they let 46 B.C. continue for 445 days so that it was later known as "The Year of Confusion." However, this brought the calendar even with the Sun so that 46 B.C. was the *last* year of confusion.

With 45 B.C. the Romans adopted a modified Egyptian calendar in which the five extra days at the end of the year were distributed throughout the year, giving us our months of uneven length. Ideally, we should have seven 30-day months and five 31-day months. Unfortunately, the Romans considered February an unlucky month and shortened it, so that we ended with a silly arrangement of seven 31-day months, four 30-day months, and one 28-day month.

In order to take care of that extra ¼ day, Caesar and Sosigenes established every fourth year with a length of 366 days. (Under the numbering of the years of the Christian era, every year divisible by 4 has the intercalary day—set as February 29. Since 1964 divided by 4 is 491, without a remainder, there is a February 29, in 1964.)

This is the "Julian year," after Julius Caesar. At the Council of Nicaea, the Christian Church adopted the Julian calendar. Christmas was finally accepted as a Church holiday after the Council of Nicaea and given a date in the Julian year. It does not, therefore, bounce about from year to year as Easter does.

The 365-day year is just 52 weeks and 1 day long. This means that if February 6, for instance, is on a Sunday in one year, it is on a Monday the next year, on a Tuesday the year after, and so on. If there were only 365-day years, then any given date would move through the days of the week in steady progression. If a 366-day year is involved, however, that year is 52 weeks and 2 days long, and if February 6 is on Tuesday that year, it is on Thursday the year after. The day has leaped over Wednesday. It is for that reason that the 366-day year is called "leap year" and February 29 is "leap day."

All would have been well if the tropical year were really exactly 365.25 days long; but it isn't. The tropical year is 365 days, 5 hours, 48 minutes, 46 seconds, or 365.24220 days long. The Julian year is, on the average, 11 minutes 14 seconds, or 0.0078 days, too long.

This may not seem much, but it means that the Julian year gains a full day on the tropical year in 128 years. As the Julian year gains, the vernal equinox, falling behind, comes earlier and earlier in the year. At the Council of Nicaea in A.D. 325, the vernal equinox was on March 21. By A.D. 453 it was on March 20, by A.D. 581 on March 19, and so on. By A.D. 1263, in the lifetime of Roger Bacon, the Julian year had gained eight days on the Sun and the vernal equinox was on March 13.

Still not fatal, but the Church looked forward to an indefinite future and Easter was tied to a vernal equinox at March 21. If this were allowed to go on, Easter would come to be celebrated in midsummer, while Christmas would edge into the spring. In 1263, therefore, Roger Bacon wrote a letter to Pope Urban IV explaining the situation. The Church, however, took over three centuries to consider the matter.

By 1582 the Julian calendar had gained two more days and the vernal equinox was falling on March 11. Pope Gregory XIII finally took action. First, he dropped ten days, changing October 5, 1582 to October 15, 1582.

That brought the calendar even with the Sun and the vernal equinox in 1583 fell on March 21 as the Council of Nicaea had decided it should.

The next step was to prevent the calendar from getting out of step again. Since the Julian year gains a full day every 128 years, it gains three full days in 384 years or, to approximate slightly, three full days in four centuries. That means that every 400 years, three leap years (according to the Julian system) ought to be omitted.

Consider the century years—1500, 1600, 1700, and so on. In the Julian year, all century years are divisible by 4 and are therefore leap years. Every 400 years there are 4 such century years, so why not keep 3 of them ordinary years, and allow only one of them (the one that is divisible by 400) to be a leap year? This arrangement will match the year more closely to the Sun and give us the "Gregorian calendar."

To summarize: Every 400 years, the Julian calendar allows 100 leap years for a total of 146,100 days. In that same 400 years, the Gregorian calendar allows only 97 leap years for a total of 146,097 days. Compare these lengths with that of 400 tropical years, which comes to 146,096.88. Whereas, in that stretch of time, the Julian year had gained 3.12 days on the Sun, the Gregorian year had gained only 0.12 days.

Still, 0.12 days in nearly 3 hours, and this means that in 3400 years the Gregorian calendar will have gained a full day on the Sun. Around A.D. 5000 we will have to consider dropping out one extra leap year.

But the Church had waited a little too long to take action. Had it done the job a century earlier, all western Europe would have changed calendars without trouble. By A.D. 1582, however, much of northern Europe had turned Protestant. These nations would far sooner remain out of step with the Sun in accordance with the dictates of the pagan Caesar, than consent to be corrected by the Pope. Therefore they kept the Julian year.

The year 1600 introduced no crisis. It was a century year but one that was divisible by 400. Therefore, it was a leap year by both the Julian and Gregorian calendars. But 1700 was a different matter. The Julian calendar had it as a leap year and the Gregorian did not. By March 1, 1700, the Julian calendar was going to be an additional day ahead of the Sun (eleven days altogether). Denmark, the Netherlands, and Protestant Germany gave in and adopted the Gregorian calendar.

Great Britain and the American colonies held out until 1752 before giving in. Because of the additional day gained in 1700, they had to drop eleven days and changed September 2, 1752 to September 13, 1752. There were riots all over England as a result, for many people came quickly to the conclusion that they had suddenly been made eleven days older by legislation.

"Give us back our eleven days!" they cried in despair.

(A more rational objection was the fact that although the third quarter of 1752 was short eleven days, landlords calmly charged a full quarter's rent.)

As a result of this, it turns out that Washington was not born on "Washington's birthday." He was born on February 22, 1732 on the Gregorian calendar, to be sure, but the date recorded in the family Bible had to be the Julian date, February 11, 1732. When the changeover took place, Washington—a remarkably sensible man—changed the date of his birthday and thus preserved the actual day.

The Eastern Orthodox nations of Europe were more stubborn than the Protestant nations. The years 1800 and 1900 went by. Both were leap years by the Julian calendar, but not by the Gregorian calendar. By 1900, then, the Julian vernal equinox was on March 8 and the Julian calendar was 13 days ahead of the Sun. It was not until after World War I that the Soviet Union, for instance, adopted the Gregorian calendar. (In doing so, the Soviets made a slight modification of the leap year pattern which made matters even more accurate. The Soviet calendar will not gain a day on the Sun until fully 35,000 years pass.)

The Orthodox churches themselves, however, *still* cling to the Julian year, which is why the Orthodox Christmas falls on January 6 on our calendar. It is still December 25 by their calendar.

In fact, a horrible thought occurs to me—

I was myself born at a time when the Julian calendar was still in force in the—ahem—old country.* Unlike George Washington, I never changed the birthdate and, as a result, each year I celebrate my birthday 13 days earlier than I should, making myself 13 days older than I have to be.

And this 13-day older me is in all the records and I can't ever change it back.

Give me back my 13 days! Give me back my 13 days! Give me back . . .

* Well, the Soviet Union, if you must know. I came here at the age of 3.

2. BEGIN AT THE BEGINNING

Each year, another New Year's Day falls upon us; and because my birthday follows hard upon New Year's Day, the beginning of the year is always a doubled occasion for great and somber soul-searching on my part.

Perhaps I can make my consciousness of passing time less poignant by thinking more objectively. For instance, who says the year starts on New Year's Day? What is there about New Year's Day that is different from any other day? What makes January 1 so special?

In fact, when we chop up time into any kind of units, how do we decide with which unit to start?

For instance, let's begin at the beginning (as I dearly love to do) and consider the day itself.

The day is composed of two parts, the daytime* and the night. Each, separately, has a natural astronomic beginning. The daytime begins with sunrise; the night begins with sunset. (Dawn and twilight encroach upon the night but that is a mere detail.)

In the latitudes in which most of humanity live, however, both daytime and night change in length during the year (one growing longer as the other grows shorter) and

* It is very annoying that "day" means both the sunlit portion of time and the twenty-four-hour period of daytime and night together. This is a completely unnecessary shortcoming of the admirable English language. I understand that the Greek language contains separate words for the two entities. I shall use "daytime" for the sunlit period and "day" for the twenty-four-hour period.

there is, therefore, a certain convenience in using daytime plus night as a single twenty-four-hour unit of time. The combination of the two, the day, is of nearly constant duration.

Well, then, should the day start at sunrise or at sunset? You might argue for the first, since in a primitive society that is when the workday begins. On the other hand, in that same society sunset is when the workday ends, and surely an ending means a new beginning.

Some groups made one decision and some the other. The Egyptians, for instance, began the day at sunrise, while the Hebrews began it at sunset.

The latter state of affairs is reflected in the very first chapter of Genesis in which the days of creation are described. In Genesis 1:5 it is written: "And the evening and the morning were the first day." Evening (that is, night) comes ahead of morning (that is, daytime) because the day starts at sunset.

This arrangement is maintained in Judaism to this day, and Jewish holidays still begin "the evening before." Christianity began as an offshoot of Judaism and remnants of this sunset beginning cling even now to some non-Jewish holidays.

The expression Christmas Eve, if taken literally, is the evening of December 25, but as we all know it really means the evening of December 24—which it would naturally mean if Christmas began "the evening before" as a Jewish holiday would. The same goes for New Year's Eve.

Another familiar example is All Hallow's Eve, the evening of the day before All Hallows' Day, which is given over to the commemoration of all the "hallows" (or "saints"). All Hallows' Day is on November 1, and All Hallows' Eve is therefore on the evening of October 31. Need I tell you that All Hallows' Eve is better known by its familiar contracted form of "Halloween."

As a matter of fact, though, neither sunset nor sunrise is now the beginning of the day. The period from sunrise to sunrise is slightly more than 24 hours for half the year as the daytime periods grow shorter, and slightly less than 24 hours for the remaining half of the year as the daytime periods grow longer. This is also true for the period from sunset to sunset.

Sunrise and sunset change in opposite directions, either approaching each other or receding from each other, so that the middle of daytime (midday) and the middle of night (midnight) remain fixed at 24-hour intervals throughout the year. (Actually, there are minor deviations but these can be ignored.)

One can begin the day at midday and count on a steady 24-hour cycle, but then the working period is split between two different dates. Far better to start the day at midnight when all decent people are asleep; and that, in fact, is what we do.

Astronomers, who are among the indecent minority not in bed asleep at midnight, long insisted on starting their day at midday so as not to break up a night's observation into two separate dates. However, the spirit of conformity was not to be withstood, and in 1925, they accepted the inconvenience of a beginning at midnight in order to get into step with the rest of the world.

All the units of time that are shorter than a day depend on the day and offer no problem. You start counting the hours from the beginning of the day; you start counting the minutes from the beginning of the hour, and so on.

Of course, when the start of the day changed its position, that affected the counting of the hours. Originally, the daytime and the night were each divided into twelve hours, beginning at, respectively, sunrise and sunset. The hours changed length with the change in length of daytime and night so that in June (in the northern hemisphere) the daytime was made up of twelve long hours and the night of twelve short hours, while in December the situation was reversed.

This manner of counting the hours still survives in the Catholic Church as "canonical hours." Thus, "prime" ("one") is the term for 6 A.M. "Tierce" ("three") is 9 A.M., "sext" ("six") is 12 A.M., and "none" ("nine") is 3 P.M. Notice that "none" is located in the middle of the afternoon when the day is warmest. The warmest part of the day might well be felt to be the middle of the day, and the word was somehow switched to the astronomic midday so that we call 12 A.M. "noon."

This older method of counting the hours also plays a part in one of the parables of Jesus (Matt. 20:1-16), in which laborers are hired at various times of the day, up to and including "the eleventh hour." The eleventh hour referred to in the parable is one hour before sunset when the working day ends. For that reason, "the eleventh hour" has come to mean the last moment in which something can be done. The force of the expression is lost on us, however, for we think of the eleventh hour as being either 11 A.M. or 11 P.M., and 11 A.M. is too early in the day to begin to feel panicky, while 11 P.M. is too late—we ought to be asleep by then.

The week originated in the Babylonian calendar where one day out of seven was devoted to rest. (The rationale was that it was an unlucky day.)

The Jews, captive in Babylon in the sixth century B.C., picked up the notion and established it on a religious basis, making it a day of happiness rather than of ill fortune. They explained its beginnings in Genesis 2:2 where, after the work of the six days of creation—"on the seventh day God ended his work which he had made; and he rested on the seventh day."

To those societies which accept the Bible as a book of special significance, the Jewish "sabbath" (from the Hebrew word for "rest") is thus defined as the seventh, and last, day of the week. This day is the one marked Saturday on our calendars, and Sunday, therefore, is the first day of a new week. All our calendars arrange the days in seven columns with Sunday first and Saturday seventh.

The early Christians began to attach special significance to the first day of the week. For one thing, it was the "Lord's day" since the Resurrection had taken place on a Sunday. Then, too, as time went on and Christians began to think of themselves as something more than a Jewish sect, it became important to them to have distinct rituals of their own. In Christian societies, therefore, Sunday, and not Saturday, became the day of rest. (Of course, in our modern effete times, Saturday and Sunday are *both* days of rest, and are lumped together as the "weekend," a period celebrated by automobile accidents.)

The fact that the work week begins on Monday causes a great many people to think of that as the first day of the week, and leads to the following children's puzzle (which I mention only because it trapped me neatly the first time I heard it).

You ask your victim to pronounce t-o, t-o-o, and t-w-o, one at a time, thinking deeply between questions. In each case he says (wondering what's up) "tooooo."

Then you say, "Now pronounce the second day of the week" and his face clears up, for he thinks he sees the trap. He is sure you are hoping he will say "tooooosday" like a lowbrow. With exaggerated precision, therefore, he says "tyoosday."

At which you look gently puzzled and say, "Isn't that strange? I always pronounce it Monday."

The month, being tied to the Moon, began, in ancient times, at a fixed phase. In theory, any phase will do. The month can start at each full Moon, or each first quarter, and so on. Actually, the most logical way is to begin each month with the new Moon—that is, on that evening when the first sliver of the growing crescent makes itself visible immediately after sunset. To any logical primitive, a new Moon is clearly being created at that time and the month should start then.

Nowadays, however, the month is freed of the Moon and is tied to the year, which is in turn based on the Sun. In our calendar, in ordinary years, the first month begins on the first day of the year, the second month on the 32nd day of the year, the third month on the 60th day of the year, the fourth month on the 91st day of the year, and so on—quite regardless of the phases of the Moon. (In a leap year, all the months from the third onward start a day late because of the existence of February 29.

But that brings us to the year. When does that begin and why?

Primitive agricultural societies must have been first aware of the year as a succession of seasons. Spring, summer, autumn, and winter were the morning, midday, evening and night of the year and, as in the case of the day, there seemed two equally qualified candidates for the post of beginning.

The beginning of the work year is the time of spring, when warmth returns to the earth and planting can begin. Should that not also be the beginning of the year in general? On the other hand, autumn marks the end of the work year, with the harvest (it is to be devoutly hoped) safely in hand. With the work year ended, ought not the new year begin?

With the development of astronomy, the beginning of the spring season was associated with the vernal equinox (see Chapter 4) which, on our calendar, falls on March 20, while the beginning of autumn is associated with the autumnal equinox which falls, half a year later, on September 23.

Some societies chose one equinox as the beginning and some the other. Among the Hebrews, both equinoxes came to be associated with a New Year's Day. One of these fell on the first day of the month of Nisan (which comes at about the vernal equinox). In the middle of that month comes the feast of Passover, which is thus tied to the vernal equinox.

Since, according to the Gospels, Jesus' Crucifixion and Resurrection occurred during the Passover season (the Last Supper was a Passover seder), Good Friday and Easter are also tied to the vernal equinox (see Chapter 1).

The Hebrews also celebrated a New Year's Day on the first two days of Tishri (which falls at about the autumnal equinox), and this became the more important of the two occasions. It is celebrated by Jews today as "Rosh Hashonah" ("head of the year"), the familiarly known "Jewish New Year."

A much later example of a New Year's Day in connection with the autumnal equinox came in connection with the French Revolution. On September 22, 1792, the French monarchy was abolished and a republic proclaimed. The Revolutionary idealists felt that since a new epoch in human history had begun, a new calendar was needed. They made September 22 the New Year's Day and established a new list of months. The first month was Vendémiare, so that September 22 became Vendémiare 1.

For thirteen years, Vendémiare 1 continued to be the official New Year's Day of the French Government, but the calendar never caught on outside France or even among the people inside France. In 1806 Napoleon gave up the struggle and officially reinstated the old calendar.

There are two important solar events in addition to the equinoxes. After the vernal equinox, the noonday Sun continues to rise higher and higher until it reaches a maximum height on June 21, which is the summer solstice (see Chapter 4), and this day, in consequence, has the longest daytime period of the year.

The height of the noonday Sun declines thereafter until it reaches the position of the autumnal equinox. It then continues to decline farther and farther till it reaches a minimum height on December 21, the winter solstice and the shortest daytime period of the year.

The summer solstice is not of much significance. "Midsummer Day" falls at about the summer solstice (the traditional English day is June 24). This is a time for gaiety and carefree joy, even folly. Shakespeare's *A Midsummer Night's Dream* is an example of a play devoted to the kind of not-to-be-taken-seriously fun of the season, and the phrase "midsummer madness" may have arisen similarly.

The winter solstice is a much more serious affair. The Sun is declining from day to day, and to a primitive society, not sure of the invariability of astronomical laws, it might well appear that *this* time, the Sun will continue its decline and disappear forever so that spring will never come again and all life will die.

Therefore, as the Sun's decline slowed from day to day and came to a halt and began to turn on December 21, there must have been great relief and joy which, in the end, became ritualized into a great religious festival, marked by gaiety and licentiousness.

The best-known examples of this are the several days of holiday among the Romans at this season of the year. The holiday was in honor of Saturn (an ancient Italian god of agriculture) and was therefore called the "Saturnalia." It was a time of feasting and of giving of presents; of good will to men, even to the point where slaves were given temporary freedom while their masters waited upon them. There was also a lot of drinking at Saturnalia parties.

In fact, the word "saturnalian" has come to mean dissolute, or characterized by unrestrained merriment.

There is logic, then, in beginning the year at the winter solstice which marks, so to speak, the birth of a new Sun, as the first appearance of a crescent after sunset marks the birth of a new Moon. Something like this may have been in Julius Caesar's mind when he reorganized the Roman calendar and made it solar rather than lunar (see Chapter 1).

The Romans had, traditionally, begun their year on March 15 (the "Ides of March"), which was intended to fall upon the vernal equinox originally but which, thanks to the sloppy way in which the Romans maintained their calendar, eventually moved far out of synchronization with the equinox. Caesar adjusted matters and moved the beginning of the year to January 1 instead, placing it nearly at the winter solstice.

This habit of beginning the year on or about the winter solstice did not become universal, however. In England (and the American colonies) March 25, intended to represent the vernal equinox, remained the official beginning of the year until 1752. It was only then that the January 1 beginning was adopted.

The beginning of a new Sun reflects itself in modern times in another way, too. In the days of the Roman Empire, the rising power of Christianity found its most dangerous competitor in Mithraism, a cult that was Persian in origin and was devoted to sun worship. The ritual centered about the mythological character of Mithras, who represented the Sun, and whose birth was celebrated on

December 25—about the time of the winter solstice. This was a good time for a holiday, anyway, for the Romans were used to celebrating the Saturnalia at that time of year.

Eventually, though, Christianity stole Mithraic thunder by establishing the birth of Jesus on December 25 (there is no biblical authority for this), so that the period of the winter solstice has come to mark the birth of both the Son and the Sun. There are some present-day moralists (of whom I am one) who find something unpleasantly reminiscent of the Roman Saturnalia in the modern secular celebration of Christmas.

But where do the years begin? It is certainly convenient to number the years, but where do we start the numbers? In ancient times, when the sense of history was not highly developed, it was sufficient to begin numbering the years with the accession of the local king or ruler. The numbering would begin over again with each new king. Where a city has an annually chosen magistrate, the year might not be numbered at all, but merely identified by the name of the magistrate for that year. Athens named its years by its archons.

When the Bible dates things at all, it does it in this manner. For instance, in II Kings 16:1, it is written: "In the seventeenth year of Pekah the son of Remaliah, Ahaz the son of Jotham king of Judah began to reign." (Pekah was the contemporary king of Israel.)

And in Luke 2:2, the time of the taxing, during which Jesus was born, is dated only as follows: "And this taxing was first made when Cyrenius was governor of Syria."

Unless you have accurate lists of kings and magistrates and know just how many years each was in power and how to relate the list of one region with that of another, you are in trouble, and it is for that reason that so many ancient dates are uncertain—even (as I shall soon explain) a date as important as that of the birth of Jesus.

A much better system would be to pick some important date in the past (preferably one far enough in the past so that you don't have to deal with negative-numbered years before that time) and number the years in progression thereafter, without ever starting over.

The Greeks made use of the Olympian Games for that purpose. This was celebrated every four years so that a four-year cycle was an "Olympiad." The Olympiads were numbered progressively, and the year itself was the 1st, 2nd, 3rd, or 4th year of a particular Olympiad.

This is needlessly complicated, however, and in the time following Alexander the Great something better was introduced into the Greek world. The ancient East was being fought over by Alexander's generals, and one of them, Seleucus, defeated another at Gaza. By this victory Seleucus was confirmed in his rule over a vast section of Asia. He determined to number the years from that battle, which took place in the 1st year of the 117th Olympiad. That year became Year 1 of the "Seleucid Era," and later years continued in succession as 2, 3, 4, 5, and so on. Nothing more elaborate than that.

The Seleucid Era was of unusual importance because Seleucus and his descendants ruled over Judea, which therefore adopted the system. Even after the Jews broke free of the Seleucids under the leadership of the Maccabees, they continued to use the Seleucid Era in dating their commercial transactions over the length and breadth of the ancient world. Those commercial records can be tied in with various local year-dating systems, so that many of them could be accurately synchronized as a result.

The most important year-dating system of the ancient world, however, was that of the "Roman Era." This began with the year in which Rome was founded. According to tradition, this was the 4th Year of the 6th Olympiad, which came to be considered as 1 A.U.C. (The abbreviation "A.U.C." stands for : "Anno Urbis Conditae"; that is, "The Year of the Founding of the City.")

Using the Roman Era, the Battle of Zama, in which Hannibal was finally defeated, was fought in 553 A.U.C., while Julius Caesar was assassinated in 710 A.U.C., and so on. This system gradually spread over the ancient world, as Rome waxed supreme, and lasted well into early medieval times.

The early Christians, anxious to show that biblical records antedated those of Greece and Rome, strove to begin counting at a date earlier than that of either the founding of Rome or the beginning of the Olympian Games. A Church historian, Eusebius of Caesarea, who lived about 1050 A.U.C., calculated that the Patriarch, Abraham, had been born 1263 years before the founding of Rome. Therefore he adopted that year as his Year 1, so that 1050 A.U.C. became 2313, Era of Abraham.

Once the Bible was thoroughly established as *the* book of the western world, it was possible to carry matters to their logical extreme and date the years from the creation of the world. The medieval Jews calculated that the creation of the world had taken place 3007 years before the founding of Rome, while various Christian calculators chose years varying from 3251 to 4755 years before the founding of Rome. These are the various "Mundane Eras" ("Eras of the World"). The Jewish Mundane Era is used today in the Jewish calendar, so that in September 1964, the Jewish year 5725 began.

The Mundane Eras have one important factor in their favor. They start early enough so that there are very few, if any, dates in recorded history that have to be given negative numbers. This is not true of the Roman Era, for instance. The founding of the Olympian Games, the Trojan War, the reign of David, the building of the Pyramids, all came before the founding of Rome and have to be given negative year numbers.

The Romans wouldn't have cared, of course, for none of the ancients were very chronology conscious, but modern historians would. In fact, modern historians are even worse off than they would have been if the Roman Era had been retained.

About 1288 A.U.C., a Syrian monk named Dionysius Exiguus, working from biblical data and secular records, calculated that Jesus must have been born in 754 A.U.C. This seemed a good time to use a beginning for counting the years, and in the time of Charlemagne (two and a half centuries after Dionysius) this notion won out.

The year 754 A.U.C. became A.D. 1 (standing for *Anno Domini,* meaning "the year of the Lord"). By this new "Christian Era," the founding of Rome took place in 753 B.C. ("before Christ"). The first year of the first Olympiad was in 776 B.C., the first year of the Seleucid Era was in 312 B.C., and so on.

This is the system used today, and means that all of ancient history from Sumer to Augustus must be dated in negative numbers, and we must forever remember that Caesar was assassinated in 44 B.C. and that the next year is number 43 and not 45.

Worse still, Dionysius was wrong in his calculations. Matthew 2:1 clearly states that "Jesus was born in Bethlehem of Judea in the days of Herod the king." This Herod is the so-called Herod the Great, who was born about 681 A.U.C., and was made king of Judea by Mark Antony in 714 A.U.C. He died (and this is known as certainly as any ancient date is known) in 750 A.U.C., and therefore Jesus could not have been born any later than 750 A.U.C.

But 750 A.U.C., according to the system of Dionysius Exiguus, is 4 B.C., and therefore you constantly find in lists of dates that Jesus was born in 4 B.C.; that is, four years before the birth of Jesus.

In fact, there is no reason to be sure that Jesus was born in the very year that Herod died. In Matthew 2:16, it is written that Herod, in an attempt to kill Jesus, ordered all male children of two years and under to be slain. This verse can be interpreted as indicating that Jesus may have been at least two years old while Herod was still alive, and might therefore have been born as early as 6 B.C. Indeed, some estimates have placed the birth of Jesus as early as 17 B.C.

Which forces me to admit sadly that although I love to begin at the beginning, I can't always be sure where the beginning is.

3. GHOST LINES IN THE SKY

My son is bearing, with strained patience, the quasi-humorous changes being rung upon his last name by his grade-school classmates. My explanation to him that the name "Asimov," properly pronounced, has a noble resonance like the distant clash of sword on shield in the age of chivalry, leaves him unmoved. The hostile look in his eyes tells me quite plainly that he considers it my duty as a father to change my name to "Smith" forthwith.

Of course, I sympathize with him, for in my time, I, too, have been victimized in this fashion. The ordinary misspellings of the uninformed I lay to one side. However, there was one time . . .

It was when I was in the Army and working out my stint in basic training. One of the courses to which we were exposed was map-reading, which had the great advantage of being better than drilling and hiking. And then, like a bolt of lightning, the sergeant in charge pronounced the fatal word "azimuth" and all faces turned toward me.

I stared back at those stalwart soldier-boys in horror, for I realized that behind every pair of beady little eyes, a small brain had suddenly discovered a source of infinite fun.

You're right. For what seemed months, I was Isaac Azimuth to every comic on the post, and every soldier on the post considered himself a comic. But, as I told myself (paraphrasing a great American poet), "This is the army, Mr. Azimuth."

Somehow, I survived.

And, as fitting revenge, what better than to tell all you inoffensive Gentle Readers, in full and leisurely detail, exactly what azimuth is?

It all starts with direction. The first, most primitive, and most useful way of indicating direction is to point. "They went that-a-way." Or, you can make use of some landmark known to one and all, "Let's head them off at the gulch."

This is all right if you are concerned with a small section of the Earth's surface; one with which you and your friends are intimately familiar. Once the horizons widen, however, there is a search for methods of giving directions that do not depend in any way on local terrain, but are the same everywhere on the Earth.

An obvious method is to make use of the direction of the rising Sun and that of the setting Sun. (These directions change from day to day, but you can take the average over the period of a year.) These are opposite directions, of course, which we call "east" and "west." Another pair of opposites can be set up perpendicular to these and be called "north" and "south."

If, at any place, north, east, south, and west are determined (and this could be done accurately enough, even in prehistoric times, by careful observations of the Sun) there is nothing, in principle, to prevent still finer directions from being established. We can have northeast, north-northeast, northeast by north, and so on.

With a compass you can accept directions of this sort, follow them for specified distances or via specified landmarks, and go wherever you are told to go. Furthermore, if you want to map the Earth, you can start at some point, travel a known distance in a known direction to another point, and locate that point (to scale) on the map. You can then do the same for a third point, and a fourth, and a fifth, and so on. In principle the entire surface of the planet can be laid out in this manner, as accurately as you wish, upon a globe.

However, the fact that a thing can be done "in principle" is cold comfort if it is unbearably tedious and would take a million men a million years. Besides, the compass was unknown to western man until the thirteenth century, and the Greek geographers, in trying to map the world, had to use other dodges.

One method was to note the position of the Sun at midday; that is at the moment just halfway between sunrise and sunset. On any particular day there will be some spots on Earth where the Sun will be directly overhead at midday. The ancient Greeks knew this to be true of southern Egypt in late June, for instance. In Europe, however, the sun at midday always fell short of the overhead point.

This could easily be explained once it was realized that the Earth was a sphere. It could furthermore be shown without difficulty that all points on Earth at which the Sun, on some particular day, fell equally short of the overhead point at midday, were on a single east-west line. Such a line could be drawn on the map and used as a reference for the location of other points. The first to do so was a Greek geographer named Dicaearchus, who lived about 300 B.C. and was one of Aristotle's pupils.

Such a line is called a line of "latitude," from a Latin word meaning broad or wide, for when making use of the usual convention of putting north at the top of a map, the east-west lines run in the direction of its width.

Naturally, a number of different lines of latitude can be determined. All run east-west and all circle the sphere of the Earth at constant distances from each other, and so are parallel. They are therefore referred to as "parallels of latitude."

The nearer the parallels of latitude to either pole, the smaller the circles they make. (If you have a globe, look at it and see.) The longest parallel is equidistant from the poles and makes the largest circle, taking in the maximum girth of the Earth. Since it divides the Earth into two equal halves, north and south, it is called the "equator" (from a Latin word meaning "equalizer").

If the Earth were cut through at the equator, the section would pass through the center of the Earth. That makes the equator a "great circle." Every sphere has an infinite number of great circles, but the equator is the only parallel of latitude that is one of them.

It early became customary to measure off the parallels of latitude in degrees. There are 360 degrees, by convention, into which the full circumference of a sphere can be divided. If you travel from the equator to the North Pole, you cover a quarter of the Earth's circumference and therefore pass over 90 degrees. Consequently, the parallels range from 0° at the equator to 90° at the North Pole (the small ° representing "degrees").

If you continue to move around the Earth past the North Pole so as to travel toward the equator again, you must pass the parallels of latitude (each of which encircles the Earth east-west) in reverse order, traveling from 90° back to 0° at the equator (but at a point directly opposite that of the equatorial beginning). Past the equator, you move across a second set of parallels circling the southern half of the globe, up to 90° at the South Pole and then back to 0°, finally at the starting point on the equator.

To differentiate the 0° to 90° stretch from equator to North Pole and the similar stretch from equator to South Pole, we speak of "north latitude" and "south latitude." Thus, Philadelphia, Pennsylvania is on the 40° north latitude parallel, while Valdivia, Chile is on the 40° south latitude parallel.

Parallels of latitude, though excellent as references about which to build a map, cannot by themselves be used to locate points on the Earth's surface. To say that Quito, Ecuador is on the equator merely tells you that it is somewhere along a circle 25,000 miles in circumference.

For accurate location one needs a gridwork of lines—a set of north-south lines as well as east-west ones. These north-south lines, running up and down the conventionally oriented map (longways) would naturally be called "longitude."

Whenever it is midday upon some spot of the Earth it is midday at all spots on the same north-south line, as one can easily show if the Earth is considered to be a rotating sphere. The north-south line is therefore a "meridian" (a corruption of a Latin word for "midday"), and we speak of "meridians of longitude."

Each meridian extends due north and south, reaching the North Pole at one extreme and the South Pole at the other. All the meridians therefore converge at both poles and are spaced most widely apart at the equator, for all the world like the boundary lines of the segments of a tangerine. If one imagines the Earth sliced in two along any meridian, the slice always cuts through the Earth's center, so that *all* meridians are great circles, and each stretches around the world a distance of approximately 25,000 miles.

By 200 B.C. maps being prepared by Greeks were marked off with both longitude and latitude. However, making the gridwork *accurate* was another thing. Latitude was all right. That merely required the determination of the average height of the midday sun or, better yet, the average height of the North Star. Such determinations could not be made as accurately in ancient Greek times as in modern times, but they could be made precisely enough to produce reasonably accurate results.

Longitude was another matter. For that you needed the *time of day*. You had to be able to compare the time at which the Sun, or better still, another star (the sun is a star) was directly above the local meridian, as compared with the time it was directly above another meridian. If a star passed over the meridian of Athens in Greece at a certain time, and over the meridian of Messina in Sicily 32 minutes later, then Messina was 8 degrees of longitude west of Athens. To determine such matters, accurate timepieces were necessary; timepieces that could be relied on to maintain synchronization to within fractions of a minute over long periods while separated by long distance; and to remain in synchronization with the Earth's rotation, too.

In ancient times, such timepieces simply did not exist and therefore even the best of the ancient geographers managed to get their meridians tangled up. Erathosthenes of Cyrene, who flourished at Alexandria in 200 B.C., thought that the meridian that passed through Alexandria also passed through Byzantium (the modern city of Istanbul, Turkey). That meridian actually passes about 70 miles east of Istanbul. Such discrepancies tended to increase in areas farther removed from home base.

Of course, once the circumference of the earth is known (and Eratosthenes himself calculated it), it is possible to calculate the east-west distance between degrees of longitude. For instance, at the equator, one degree of longitude is equal to about 69.5 miles, while at a latitude of 40° (either north or south of the equator), it is only about 53.2 miles, and so on. However, accurate measurements of distance over mountainous territory or, worse yet, over stretches of open ocean, are quite difficult.

In early modern times, when European nations first began to make long ocean voyages, this became a horrible problem. Sea captains never knew certainly where they were, and making port was a matter of praying as well as sailing. In 1598 Spain, then still a major seagoing nation, offered a reward for anyone who would devise a timepiece that could be used on board ship, but the reward went begging.

In 1656 the Dutch astronomer Christian Huygens invented the pendulum clock—the first accurate timepiece. It could be used only on land, however. The pitching, rolling, and yawing of a ship put the pendulum off its feed at once.

Great Britain was a major maritime nation after 1600, and in 1675 Charles II founded the observatory in Greenwich (then a London suburb, now part of Greater London) for the express purpose of carrying through the necessary astronomical observations that would make the accurate determination of longitude possible.

But a good timepiece was still needed, and in 1714 the British Government offered a large fortune (in those days) of 20,000 pounds for anyone who could devise a good clock that would work on shipboard.

The problem was tackled by John Harrison, a Yorkshire mechanic, self-trained and gifted with mechanical genius. Beginning in 1728 he built a series of five clocks, each better than the one before. Each was so mounted that it could take the sway of a ship without being affected. Each was more accurate at sea than other clocks of the time were on land. One of them was off by less than a minute after five months at sea. Harrison's first clocks were perhaps too large and heavy to be completely practical, but the fifth was no bigger than a large watch.

The British Parliament put on an extraordinary display of meanness in this connection, for it wore Harrison out in its continual delays in paying him the money he had earned and in demanding more and ever more models and tests. (Possibly this was because Harrison was a provincial mechanic and not a gentleman scientist of the Royal Society.) However, King George III himself took a personal interest in the case and backed Harrison, who finally received his money in 1765, by which time he was over 70 years old.

It is only in the last two hundred years, then, that the latitude-longitude gridwork on the earth became really accurate.

Even after precise longitude determinations became possible, a problem remained. There is no natural reference base for longitude; nothing like the equator in the case of latitude. Different nations therefore used different systems, usually basing "zero longitude" on the meridian passing through the local capital. The use of different systems was confusing and the risk was run of rescue operations at sea being hampered, to say nothing of war maneuvers among allies being stymied.

To settle matters, the important maritime nations of the world gathered in Washington, D.C. in 1884 and held the "Washington Meridian Conference." The logical decision was reached to let the Greenwich observatory serve as base since Great Britain was at the very height of its maritime power. The meridian passing through Greenwich is, therefore, the "prime meridian" and has a longitude of 0°.

The degrees of longitude are then marked off to the west and east as "west longitude" and "east longitude." The two meet again at the opposite side of the world from the prime meridian. There we have the 180° meridian which runs down the middle of the Pacific Ocean.

Every degree of latitude (or longitude) is broken up

into 60 minutes (′), every minute into 60 seconds (″), while the seconds can be broken up into tenths, hundredths, and so on. Every point on the earth can be located uniquely by means of latitude and longitude. For instance, an agreed-upon reference point within New York City is at 40° 45′ 06″ north latitude and 73° 59′ 39″ west longitude; while Los Angeles is at 34° 03′ 15″ north latitude and 118° 14′ 28″ west longitude.

The North Pole and the South Pole have no longitude, for all the meridians converge there. The North Pole is defined by latitude alone, for 90° north latitude represents one single point—the North Pole. Similarly, 90° south latitude represents the single point of the South Pole.

It is possible to locate longitude in terms of time rather than in terms of degrees. The complete day of 24 hours is spread around the 360° of longitude. This means that if two places differ by 15° in longitude, they also differ by 1 hour in local time. If it is exactly noon on the prime meridian, it is 1 P.M. at 15° east longitude and 11 A.M. at 15° west longitude.

If we decide to call prime meridian 0:00:00 we can assign west longitude positive time readings and east longitude negative time readings. All points on 15° west longitude become +1:00:00 and all points on 15° east longitude become −1:00:00.

Since New York City is at 73° 59′ 39″ west longitude it is 4 hours 55 minutes 59 seconds earlier than London and can therefore be located at +4:55:59. Similarly, Los Angeles, still farther west, is at +8:04:48.

In short, every point on Earth, except for the poles, can be located by a latitude and a time. The North and South Poles have latitude only and no local times, since they have no meridians. This does not mean, of course, that there is no time at the poles; only that the system for measuring local times, which works elsewhere on Earth, breaks down at the poles. Other systems can be used there; one pole might be assigned Greenwich time, for instance, while the other is assigned the time of the 180° meridian.

In the ordinary mapping of the globe, both latitude and

longitude are given in ordinary degrees. However, the time system for longitude is used to establish local time zones over the face of the Earth, and the 180° meridian becomes the "International Date Line" (slightly bent for geographical convenience). All sorts of interesting paradoxes become possible, but that is for another article another day.

And what about mapping the sky? This concerned astronomers even before the problem of the mapping of the Earth, really, for whereas only small portions of the Earth are visible to any one man at any one time, the entire expanse of half a sphere is visible overhead.

The "celestial sphere" is most easily mapped as an extension of the earthly sphere. If the axis of the Earth is imagined extended through space until it cuts the celestial sphere, the intersection would come at the "North Celestial Pole" and the "South Celestial Pole." ("Celestial," by the way, is from a Latin word for "sky.")

The celestial sphere seems to rotate east to west about the Earth's axis as a reflection of the actual rotation of the Earth west to east about that axis. Therefore, the North Celestial Pole and the South Celestial Pole are fixed points that do not partake in the celestial rotation, just as the North Pole and the South Pole do not partake in the earthly rotation.

The near neighborhood of the North Celestial Pole is marked by a bright star, Polaris, also called the "pole star" and the "north star," which is only a degree or so from it and makes a small circle about it each day. The circle is so small that the star seems fixed in position day after day, year after year, and can be used as a reference point to determine north, and therefore all other directions. Its importance to travel in the days before the compass was incalculable.

The imaginary reference lines on the Earth can all be transferred by projection to the sky, so that the sky, like the Earth, can be covered with a gridwork of ghost lines. There would be the "celestial equator," making up a great circle equidistant from the celestial poles; and "celestial latitude" and "celestial longitude" also.

The celestial latitude is called "declination," and is measured in degrees. The northern half of the celestial sphere ("north celestial latitude") has its declination given as a positive value; the southern half ("south celestial latitude") as a negative value. Thus, Polaris has a declination of roughly +89°; Pollux one of about +30°; Sirius one of about −15°; and Acrux (the brightest star of the Southern Cross) a declination of about −60°.

The celestial longitude is called "right ascension" and the sky has a prime meridian of its own that is less arbitrary than the one on Earth, one which could therefore be set and agreed upon quite early in the game.

The plane of the Earth's orbit about the Sun cuts the celestial sphere in a great circle called the "ecliptic" (see Chapter 4). The Sun seems to move exactly along the line of the ecliptic, in other words.

Because the Earth's axis is tipped to the plane of Earth's orbit by 23.5°, the two great circles of the ecliptic and the celestial equator are angled to one another by that same 23.5°.

The ecliptic crosses the celestial equator at two points. When the Sun is at either point, the day and night are equal in length (twelve hours each) all over the Earth. Those points are therefore the "equinoxes," from Latin words meaning "equal nights."

At one of these points the Sun is moving from negative to positive declination, and that is the "vernal equinox" because it occurs on March 20 and marks the beginning of spring in the Northern Hemisphere, where most of mankind lives. At the other point the Sun is moving from positive to negative declination and that is the autumnal equinox, falling on September 23, the beginning of the northern autumn.

The point of the vernal equinox falls on a celestial meridian which is assigned a value of 0° right ascension. The celestial longitude is then measured eastward only (either in degrees or in hours) all the way around, until it returns to itself as 360° right ascension.

By locating a star through declination and right ascension one does precisely the same thing as locating a point on Earth through latitude and longitude.

An odd difference is this, though. The Earth's prime meridian is fixed through time, so that a point on the Earth's surface does not change its longitude from day to day. However, the Earth's axis makes a slow revolution once in 25,800 years, and because of this the celestial equator slowly shifts, and the points at which it crosses the ecliptic move slowly westward.

The vernal equinox moves westward, then, circling the sky every 25,800 years, so that each year the moment in time of the vernal equinox comes just a trifle sooner than it otherwise would. The moment precedes the theoretical time and the phenomenon is therefore called "the precession of the equinoxes."

As the vernal equinox moves westward, every point on the celestial sphere has its right ascension (measured from that vernal equinox) increase. It moves up about $1/7$ of a second of arc each day, if my calculations are correct.

This system of locating points in the sky is called the "Equatorial System" because it is based on the location of the celestial equator and the celestial poles.

A second system may be established based on the observer himself. Instead of a "North Celestial Pole" based on a rotating Earth, we can establish a point directly overhead, each person on Earth having his own overhead point—although for people over a restricted area, say that of New York City, the different overhead points are practically identical.

The overhead point is the "zenith," which is a medieval misspelling of part of an Arabic phrase meaning "overhead." The point directly opposite in that part of the celestial sphere which lies under the Earth is the "nadir," a medieval misspelling of an Arabic word meaning "opposite."

The great circle that runs around the celestial sphere, equidistant from the zenith and nadir, is the "horizon," from a Greek word meaning "boundary," because to us it seems the boundary between sky and Earth (if the Earth were perfectly level, as it is at sea). This system of locating points in the sky is therefore called the "Horizon System."

The north-south great circle traveling from horizon to horizon through the zenith is the meridian. The east-west great circle traveling from horizon to horizon through the zenith, and making a right angle with the meridian, is the "prime vertical."

A point in the sky can then be said to be so many degrees (positive) above the horizon or so many degrees (negative) below the horizon, this being the "altitude." Once that is determined, the exact point in the sky can be located by measuring on that altiude the number of degrees westward from the southern half of the meridian. At least astronomers do that. Navigators and surveyors measure the number of degrees eastward from the north end of the meridian. (In both cases the direction of measure is clockwise.)

The number of degrees west of the southern edge of the meridian (or east of the northern edge, depending on the system used) is the *azimuth*. The word is a less corrupt form of the Arabic expression from which "zenith" also comes.

If you set north as having an azimuth of 0°, then east has an azimuth of 90°, south an azimuth of 180°, and west an azimuth of 270°. Instead of boxing the compass with outlandish names you can plot direction by degrees.

And as for myself?

Why, I have an azimuth of isaac. Naturally.

4. THE HEAVENLY ZOO

On July 20, 1963 there was a total eclipse of the Sun, visible in parts of Maine, but not quite visible in its total aspect from my house. In order to see the total eclipse I would have had to drive two hundred miles, take a chance on clouds, then drive back two hundred miles, braving the traffic congestion produced by thousands of other New Englanders with the same notion.

I decided not to (as it happened, clouds interfered with seeing, so it was just as well) and caught fugitive glimpses of an eclipse that was only 95 per cent total, from my backyard. However, the difference between a 95 per cent eclipse and a 100 per cent eclipse is the difference between a notion of water and an ocean of water, so I did not feel very overwhelmed by what I saw.

What makes a total eclipse so remarkable is the sheer astronomical accident that the Moon fits so snugly over the Sun. The Moon is just large enough to cover the Sun completely (at times) so that a temporary night falls and the stars spring out. And it is just small enough so that during the Sun's obscuration, the corona, especially the brighter parts near the body of the Sun, is completely visible.

The apparent size of the Sun and Moon depends upon both their actual size and their distance from us. The diameter of the Moon is 2160 miles while that of the Sun is 864,000 miles. The ratio of the diameter of the Sun to that of the Moon is 864,000/2160 or 400. In other words, if both were at the same distance from us, the Sun would appear to be 400 times as broad as the Moon.

However, the Sun is farther away from us than the Moon is, and therefore appears smaller for its size than the Moon does. At great distances, such as those which characterize the Moon and the Sun, doubling the distance halves the apparent diameter. Remembering that, consider that the average distance of the Moon from us is 238,000 miles while that of the Sun is 93,000,000 miles. The ratio of the distance of the Sun to that of the Moon is 93,000,000/238,000 or 390. The Sun's apparent diameter is cut down in proportion.

In other words, the two effects just about cancel. The Sun's greater distance makes up for its greater size and the result is that the Moon and the Sun *appear* to be equal in size. The apparent angular diameter of the Sun averages 32 minutes of arc, while that of the Moon averages 31 minutes of arc.*

These are average values because both Moon and Earth possess elliptical orbits. The Moon is closer to the Earth (and therefore appears larger) at some times than at others, while the Earth is closer to the Sun (which therefore appears larger) at some times than at others. This variation in apparent diameter is only 3 per cent for the Sun and about 5 per cent for the Moon, so that it goes unnoticed by the casual observer.

There is no astronomical reason why Moon and Sun should fit so well. It is the sheerest of coincidence, and only the Earth among all the planets is blessed in this fashion. Indeed, if it is true, as astronomers suspect, that the Moon's distance from the Earth is gradually increasing as a result of tidal friction, then this excellent fit even here on Earth is only true of our own geologic era. The Moon was too large for an ideal total eclipse in the far past and will be too small for any total eclipse at all in the far future.

Of course, there is a price to pay for this excellent fit. The fact that the Moon and Sun are roughly equal in apparent diameter means that the conical shadow of the

* One degree equals 60 minutes, so that both Sun and Moon are about half a degree in diameter.

Moon comes to a vanishing point near the Earth's surface. If the two bodies were exactly equal in apparent size the shadow would come to a pointed end exactly at the Earth's surface, and the eclipse would be total for only an instant of time. In other words, as the Moon covered the last sliver of Sun (and kept on moving, of course) the first sliver of Sun would begin to appear on the other side.

Under the most favorable conditions, when the Moon is as close as possible (and therefore as apparently large as possible) while the Sun is as far as possible (and therefore as apparently small as possible), the Moon's shadow comes to a point well below the Earth's surface and we pass through a measurable thickness of that shadow. In other words, after the unusually large Moon covers the last sliver of the unusually small Sun, it continues to move for a short interval of time before it ceases to overlap the Sun and allows the first sliver of it to appear at the other side. An eclipse, under the most favorable conditions, can be 7½ minutes long.

On the other hand, if the Moon is smaller than average in appearance, and the Sun larger, the Moon's shadow will fall short of the Earth's surface altogether. The small Moon will not completely cover the larger Sun, even when both are centered in the sky. Instead, a thin ring of Sun will appear all around the Moon. This is an "annular eclipse" (from a Latin word for "ring"). Since the Moon's apparent diameter averages somewhat less than the Sun's, annular eclipses are a bit more likely than total eclipses.

This situation scarcely allows astronomers (and ordinary beauty-loving mortals, too) to get a good look, since not only does a total eclipse of the Sun last for only a few minutes, but it can be seen only over that small portion of the Earth's surface which is intersected by the narrow shadow of the Moon.

To make matters worse, we don't even get as many eclipses as we might. An eclipse of the Sun occurs whenever the Moon gets between ourselves and the Sun. But that happens at every new Moon; in fact the Moon is "new" because it is between us and the Sun so that it is the opposite side (the one we don't see) that is sunlit, and we only get, at best, the sight of a very thin crescent sliver of light at one edge of the Moon. Well, since there are twelve new Moons each year (sometimes thirteen) we ought to see twelve eclipses of the Sun each year, and sometimes thirteen. No?

No! At most we see five eclipses of the Sun each year (all at widely separated portions of the Earth's surface, of course) and sometimes as few as two. What happens the rest of the time? Let's see.

The Earth's orbit about the Sun is all in one plane. That is, you can draw an absolutely flat sheet through the entire orbit. The Sun itself will be located in this plane as well. (This is no coincidence. The law of gravity makes it necessary.)

If we imagine this plane of the Earth's orbit carried out infinitely to the stars, we, standing on the Earth's surface, will see that plane cutting the celestial sphere into two equal halves. The line of intersection will form a "great circle" about the sky, and this line is called the "ecliptic."

Of course, it is an imaginary line and not visible to the eye. Nevertheless, it can be located if we use the Sun as a marker. Since the plane of the Earth's orbit passes through the Sun, we are sighting along the plane when we look at the Sun. The Sun's position in the sky always falls upon the line of the ecliptic. Therefore, in order to mark out the ecliptic against the starry background, we need only follow the apparent path of the Sun through the sky. (I am referring now not to the daily path from east to west, which is the reflection of Earth's rotation, but rather the path of the Sun from west to east against the starry background, which is the reflection of the Earth's revolution about the Sun.)

Of course, when the Sun is in the sky the stars are not visible, being blanked out by the scattered sunlight that turns the sky blue. How then can the position of the Sun among the stars be made out?

Well, since the Sun travels among the stars, the half of the sky which is invisible by day and the half which is visible by night shifts a bit from day to day and from night to night. By watching the night skies throughout the year the stars can be mapped throughout the entire circuit of the ecliptic. It then becomes possible to calculate the position of the Sun against the stars on each particular day, since there is always just one position that will account for the exact appearance of the night sky on any particular night.

If you prepare a celestial sphere—that is, a globe with the stars marked out upon it—you can draw an accurate great circle upon it representing the Sun's path. The time it takes the Sun to make one complete trip about the ecliptic (in appearance) is about 365¼ days, and it is this which defines the "year."

The Moon travels about the Earth in an ellipse and there is a plane that can be drawn to include its entire orbit, this plane passing through the Earth itself. When we look at the Moon we are sighting along this plane, and the Moon marks out the intersection of the plane with the starry background. The stars may be seen even when the Moon is in the sky, so that marking out the Moon's path (also a great circle) is far easier than marking out the Sun's. The time it takes the Moon to make one complete trip about its path, about 27⅓ days, defines the "sidereal month" (see Chapter 6).

Now if the plane of the Moon's orbit about the Earth coincided with the plane of the Earth's orbit about the Sun, both Moon and Sun would mark out the same circular line against the stars. Imagine them starting from the same position in the sky. The Moon would make a complete circuit of the ecliptic in 28 days, then spend an additional day and a half catching up to the Sun, which had also been moving (though much more slowly) in the interval. Every 29½ days there would be a new Moon and an eclipse of the Sun.

Furthermore, once every 29½ days, there would be a full Moon, when the Moon was precisely on the side opposite to that of the Sun so that we would see its entire visible hemisphere lit by the Sun. But at that time the Moon should pass into the Earth's shadow and there would be a total eclipse of the Moon.

All this does not happen every 29½ days because the plane of the Moon's orbit about the Earth does *not* coincide with the plane of the Earth's orbit about the Sun. The two planes make an angle of 5°8′ (or 308 minutes of arc). The two great circles, if marked out on a celestial sphere, would be set off from each other at a slight slant. They would cross at two points, diametrically opposed, and would be separated by a maximum amount exactly halfway between the crossing point. (The crossing points are called "nodes," a Latin word meaning "knots.")

If you have trouble visualizing this, the best thing is to get a basketball and two rubber bands and try a few experiments. If you form a great circle of each rubber band (one that divides the globe into two equal halves) and make them non-coincident, you will see that they cross each other in the manner I have described.

At the points of maximum separation of the Moon's path from the ecliptic, the angular distance between them is 308 minutes of arc. This is a distance equal to roughly ten times the apparent diameter of either the Sun or the Moon. This means that if the Moon happens to overtake the Sun at a point of maximum separation, there will be enough space between them to fit in nine circles in a row, each the apparent size of Moon or Sun.

In most cases, then, the Moon, in overtaking the Sun, will pass above it or below it with plenty of room to spare, and there will be no eclipse.

Of course, if the Moon happens to overtake the Sun at a point near one of the two nodes, then the Moon does get into the way of the Sun and an eclipse takes place. This happens only, as I said, from two to five times a year. If the motions of the Sun and Moon are adequately analyzed mathematically, then it becomes easy to predict when such meetings will take place in the future, and when they have taken place in the past, and exactly from what parts of the Earth's surface the eclipse will be visible.

Thus, Herodotus tells us that the Ionian philosopher, Thales, predicted an eclipse that came just in time to stop a battle between the Lydians and the Medians. (With such a sign of divine displeasure, there was no use going on with the war.) The battle took place in Asia Minor sometime after 600 B.C., and astronomical calculations show that a total eclipse of the Sun was visible from Asia Minor on May 28, 585 B.C. This star-crossed battle, therefore, is the earliest event in history which can be dated to the exact day.

The ecliptic served early mankind another purpose besides acting as a site for eclipses. It was an eternal calendar, inscribed in the sky.

The earliest calendars were based on the circuits of the Moon, for as the Moon moves about the sky, it goes through very pronounced phase changes that even the most casual observer can't help but notice. The 29½ days it takes to go from new Moon to new Moon is the "synodic month" (see Chapter 6).

The trouble with this system is that in the countries civilized enough to have a calendar, there are important periodic phenomena (the flooding the Nile, for instance, or the coming of seasonal rains, or seasonal cold) that do not fit in well with the synodic month. There weren't a whole number of months from Nile flood to Nile flood. The average interval was somewhere between twelve and thirteen months.

In Egypt it came to be noticed that the average intervals between the floods coincided with one complete Sun-circuit (the year). The result was that calendars came to consist of years subdivided into months. In Babylonia and, by dint of copying, among the Greeks and Jews, the months were tied firmly to the Moon, so that the year was made up sometimes of 12 months and sometimes of 13 months in a complicated pattern that repeated itself every 19 years. This served to keep the years in line with the seasons and the months in line with the phases of the Moon. However, it meant that individual years were of different lengths (see Chapter 1).

The Egyptians and, by dint of copying, the Romans and ourselves abandoned the Moon and made each year equal in length, and each with 12 slightly long months. The "calendar month" averaged 30½ days long in place of the 29½ days of the synodic month. This meant the months fell out of line with the phases of the Moon, but mankind survived that.

The progress of the Sun along the ecliptic marked off the calendar, and since the year (one complete circuit) was divided into 12 months it seemed natural to divide the ecliptic into 12 sections. The Sun would travel through one section in one month, through the section to the east of that the next month, through still another section the third month, and so on. After 12 months it would come back to the first section.

Each section of the ecliptic has its own pattern of stars, and to identify one section from another it is the most natural thing in the world to use those patterns. If one section has four stars in a roughly square configuration it might be called "the square"; another section might be the "V-shape," another the "large triangle," and so on.

Unfortunately, most people don't have my neat, geometrical way of thinking and they tend to see complex figures rather than simple, clean shapes. A group of stars arranged in a V might suggest the head and horns of a bull, for instance. The Babylonians worked up such imaginative patterns for each section of the ecliptic and the Greeks borrowed these, giving each a Greek name. The Romans borrowed the list next, giving them Latin names, and passing them on to us.

The following is the list, with each name in Latin and in English: 1) Aries, the Ram; 2) Taurus, the Bull; 3) Gemini, the Twins; 4) Cancer, the Crab; 5) Leo, the Lion; 6) Virgo, the Virgin; 7) Libra, the Scales; 8) Scorpio, the Scorpion; 9) Sagittarius, the Archer; 10) Capricornus, the Goat; 11) Aquarius, the Water-Carrier; 12) Pisces, the Fishes.

As you see, seven of the constellations represent animals. An eighth, Sagittarius, is usually drawn as a centaur, which may be considered an animal, I suppose. Then, if we remember that human beings are part of the animal kingdom, the only strictly nonanimal constellation is Libra. The Greeks consequently called this band of contellations *o zodiakos kyklos* or "the circle of little animals," and this has come down to us as the Zodiac.

In fact, in the sky as a whole, modern astronomers recognize 88 constellations. Of these 30 (most of them constellations of the southern skies, invented by moderns) represent inanimate objects. Of the remaining 58, mostly ancient, 36 represent mammals (including 14 human beings), 9 represent birds, 6 represent reptiles, 4 represent fish, and 3 represent arthropods. Quite a heavenly zoo!

Odd, though, considering that most of the constellations were invented by an agricultural society, that not one represents a member of the plant kingdom. Or can that be used to argue that the early star-gazers were herdsmen and not farmers?

The line of the ecliptic is set at an angle of 23½° to the celestial equator (see Chapter 3) since, as is usually stated, the Earth's axis is tipped 23½°.

At two points, then, the ecliptic crosses the celestial equator and those two crossing points are the "equinoxes" ("equal nights"). When the Sun is at those crossing points, it shines directly over the equator and days and nights are equal (twelve hours each) the world over. Hence, the name.

One of the equinoxes is reached when the Sun, in its path along the ecliptic, moves from the southern celestial hemisphere into the northern. It is rising higher in the sky (to us in the Northern Hemisphere) and spring is on its way. That, therefore, is the "vernal equinox," and it is on March 20.

On that day (at least in ancient Greek times) the Sun entered the constellation of Aries. Since the vernal equinox is a good time to begin the year for any agricultural society, it is customary to begin the list of the constellations of the Zodiac, as I did, with Aries.

The Sun stays about one month in each constellation, so it is in Aries from March 20 to April 19, in Taurus from April 20 to May 20, and so on (at least that was the lineup in Greek times).

As the Sun continues to move along the ecliptic after the vernal equinox, it moves farther and farther north of the celestial equator, rising higher and higher in our northern skies. Finally, halfway between the two equinoxes, on June 21, it reaches the point of maximum separation between ecliptic and celestial equator. Momentarily it "stands still" in its north-south motion, then "turns" and begins (it appears to us) to travel south again. This is the time of the "summer solstice," where "solstice" is from the Latin meaning "sun stand-still."

At that time the position of the Sun is a full 23½° north of the celestial equator and it is entering the constellation of Cancer. Consequently the line of 23½° north latitude on Earth, the line over which the Sun is shining on June 20, is the "Tropic of Cancer." ("Tropic" is from a Greek word meaning "to turn.")

On September 23, the Sun has reached the "autumnal equinox" as it enters the constellation of Libra. It then moves south of the celestial equator, reaching the point of maximum southerliness on December 21, when it enters the constellation of Capricorn. This is the "winter solstice," and the line of 23½° south latitude on the Earth is (you guessed it) the "Tropic of Capricorn."

Here is a complication! The Earth's axis "wobbles." If the line of the axis were extended to the celestial sphere, each pole would draw a slow circle, 47° in diameter, as it moved. The position of the celestial equator depends on the tilt of the axis and so the celestial equator moves bodily against the background of the stars from east to west in a direction parallel to the ecliptic. The position of the equinoxes (the intersection of the moving celestial equator with the unmoving ecliptic) travels westward to meet the Sun.

The equinox completes a circuit about the ecliptic in 25,760 years, which means that in 1 year the vernal equinox moves 360/25,760 or 0.014 degrees. The Sun, in making its west-to-east circuit, comes to the vernal equinox which is 0.014 degrees west of its position at the last crossing. The Sun must travel that additional 0.014 degrees to make a truly complete circuit with respect to the stars. It takes 20 minutes of motion to cover that additional 0.014 degrees. Because the equinox precedes itself and is reached 20 minutes ahead of schedule each year, this motion of the Earth's axis is called "the precession of the equinoxes."

Because of the precession of the equinoxes, the vernal equinox moves one full constellation of the Zodiac every 2150 years. In the time of the Pyramid builders, the Sun entered Taurus at the time of the vernal equinox. In the time of the Greeks, it entered Aries. In modern times, it enters Pisces. In A.D. 4000 it will enter Aquarius.

The complete circle made by the Sun with respect to the stars takes 365 days, 6 hours, 9 minutes, 10 seconds. This is the "sidereal year." The complete circle from equinox to equinox takes 20 minutes less; 365 days, 5 hours, 48 minutes, 45 seconds. This is the "tropical year," because it also measures the time required for the Sun to move from tropic to tropic and back again.

It is the tropical year and not the sidereal year that governs our seasons, so it is the tropical year we mean when we speak of *the* year.

The scholars of ancient times noted that the position of the Sun in the Zodiac had a profound effect on the Earth. Whenever it was in Leo, for instance, the Sun shone with a lion's strength and it was invariably hot; when it was in Aquarius, the water-carrier usually tipped his urn so that there was much snow. Furthermore, eclipses were clearly meant to indicate catastrophe, since catastrophe always followed eclipses. (Catastrophes also followed lack of eclipses but no one paid attention to that.)

Naturally, scholars sought for other effects and found them in the movement of the five bright star-like objects, Mercury, Venus, Mars, Jupiter, and Saturn. These, like the Sun and Moon, moved against the starry background and all were therefore called "planetes" ("wanderers") by the Greeks. We call them "planets."

The five star-like planets circle the Sun as the Earth does and the planes of their orbits are tipped only slightly to that of the Earth. This means they seem to move in the ecliptic, as the Sun and Moon do, progressing through the constellations of the Zodiac.

Their motions, unlike those of the Sun and the Moon, are quite complicated. Because of the motion of the Earth, the tracks made by the star-like planets form loops now and then. This made it possible for the Greeks to have five centuries of fun working out wrong theories to account for those motions.

Still, though the theories might be wrong, they sufficed to work out what the planetary positions were in the past and what they would be in the future. All one had to do was to decide what particular influence was exerted by a particular planet in a particular constellation of the Zodiac; note the positions of all the planets at the time of a person's birth; and everything was set. The decision as to the particular influences presents no problem. You make any decision you care to. The pseudo-science of astrology invents such influences without any visible difficulty. Every astrologer has his own set.

To astrologers, moreover, nothing has happened since the time of the Greeks. The period from March 20 to April 19 is still governed by the "sign of Aries," even though the Sun is in Pisces at that time nowadays, thanks to the precession of the equinoxes. For that reason it is now necessary to distinguish between the "signs of the Zodiac" and the "constellations of the Zodiac." The signs *now* are what the constellations *were* two thousand years ago. I've never heard that this bothered any astrologer in the world.

All this and more occurred to me some time ago when I was invited to be on a well-known television conversation show that was scheduled to deal with the subject of astrology. I was to represent science against the other three members of the panel, all of whom were professional astrologers.

For a moment I felt that I must accept, for surely it was my duty as a rationalist to strike a blow against folly and superstition. Then other thoughts occurred to me.

The three practitioners would undoubtedly be experts at their own particular line of gobbledygook and could easily speak a gallon of nonsense while I was struggling with a half pint of reason.

Furthermore, astrologers are adept at that line of argument that all pseudo-scientists consider "evidence." The line would be something like this, "People born under Leo are leaders of men, because the lion is the king of beasts, and the proof is that Napoleon was born under the sign of Leo."

Suppose, then, I were to say, "But one-twelfth of living human beings, amounting to 250,000,000 individuals were born in Taurus. Have you, or has anybody, ever tried to determine whether the proportion of leaders among them is significantly greater than among non-Leos? And how would you test for leadership, objectively, anyway?"

Even if I managed to say all this, I would merely be stared at as a lunatic and, very likely, as a dangerous subversive. And the general public, which, in this year of 1968, ardently believes in astrology and supports more astrologers in affluence (I strongly suspect) than existed in all previous centuries combined, would arrange lynching parties.

So as I wavered between the desire to fight for the right, and the suspicion that the right would be massacred and sunk without a trace, I decided to turn to astrology for help. Surely, a bit of astrologic analysis would tell me what was in store for me in any such confrontation.

Since I was born on January 2, that placed me under the sign of Capricornus—the goat.

That did it! Politely, but very firmly, I refused to be on the program!

5. ROLL CALL

When all the world was young (and I was a teen-ager), one way to give a science fiction story a good title was to make use of the name of some heavenly body. Among my own first few science fiction stories, for instance, were such items as "Marooned off Vesta," "Christmas on Ganymede," and "The Callistan Menace." (Real swinging titles, man!)

This has gone out of fashion, alas, but the fact remains that in the 1930's, a whole generation of science fiction fans grew up with the names of the bodies of the Solar System as familiar to them as the names of the American states. Ten to one they didn't know why the names were what they were, or how they came to be applied to the bodies of the Solar System or even, in some cases, how they were pronounced—but who cared? When a tentacled monster came from Umbriel or Io, how much more impressive that was than if it had merely come from Philadelphia.

But ignorance must be battled. Let us, therefore, take up the matter of the names, call the roll of the Solar System in the order (more or less) in which the names were applied, and see what sense can be made of them.

The *Earth* itself should come first, I suppose. Earth is an old Teutonic word, but it is one of the glories of the English language that we always turn to the classic tongues as well. The Greek word for Earth was *Gaia* or, in Latin spelling, *Gaea*. This gives us "geography" ("earth-writing"), "geology" ("earth-discourse"), "geometry" ("earth-measure"), and so on.

The Latin word is *Terra.* In science fiction stories a human being from Earth may be an "Earthling" or an "Earthman," but he is frequently a "Terrestrial" while a creature from another world is almost invariably an "extra-Terrestrial."

The Romans also referred to the Earth as *Tellus Mater* ("Mother Earth" is what it means). The genitive form of *tellus* is *telluris,* so Earthmen are occasionally referred to in s.f. stories as "Tellurians." There is also a chemical element "tellurium," named in honor of this version of the name of our planet.

But putting Earth to one side, the first two heavenly bodies to have been noticed were, undoubtedly and obviously, the *Sun* and the *Moon*, which, like Earth, are old Teutonic words.

To the Greeks the Sun was *Helios,* and to the Romans it was *Sol.* For ourselves, Helios is almost gone, although we have "helium" as the name of an element originally found in the Sun, "heliotrope" ("sun-turn") for the sunflower, and so on.

Sol persists better. The common adjective derived from "sun" may be "sunny," but the scholarly one is "solar." We may speak of a sunny day and a sunny disposition, but never of the "Sunny System." It is always the "Solar System." In science fiction, the Sun is often spoken of as Sol, and the Earth may even be referred to as "Sol III."

The Greek word for the Moon is *Selene,* and the Latin word is *Luna.* The first lingers on in the name of the chemical element "selenium," which was named for the Moon. And the study of the Moon's surface features may be called "selenography." The Latin name appears in the common adjective, however, so that one speaks of a "lunar crescent" or a "lunar eclipse." Also, because of the theory that exposure to the light of the full Moon drove men crazy ("moon-struck"), we obtained the word "lunatic."

I have a theory that the notion of naming the heavenly bodies after mythological characters did not originate with the Greeks, but that it was a deliberate piece of copy-cattishness.

To be sure, one speaks of *Helios* as the god of the Sun and *Gaea* as the goddess of the Earth, but it seems obvious to me that the words came first, to express the physical objects, and that these were personified into gods and goddesses later on.

The later Greeks did, in fact, feel this lack of mythological character and tried to make Apollo the god of the Sun and Artemis (Diana to the Romans) the goddess of the Moon. This may have taken hold of the Greek scholars but not of the ordinary folk, for whom Sun and Moon remained *Helios* and *Selene*. (Nevertheless, the influence of this Greek attempt on later scholars was such that no other important heavenly body was named for Apollo and Artemis.)

I would like to clinch this theory of mine, now, by taking up another heavenly body.

After the Sun and Moon, the next bodies to be recognized as important individual entities must surely have been the five bright "stars" whose positions with respect to the real stars were not fixed and which therefore, along with the Sun and the Moon, were called planets (see Chapter 4).

The brightest of these "stars" is the one we call Venus, and it must have been the first one noticed—but not necessarily as an individual. Venus sometimes appears in the evening after sunset, and sometimes in the morning before sunrise, depending on which part of its orbit it happens to occupy. It is therefore the "Evening Star" sometimes and the "Morning Star" at other times. To the early Greeks, these seemed two separate objects and each was given a name.

The Evening Star, which always appeared in the west near the setting Sun, was named *Hesperos* ("evening" or "west"). The equivalent Latin name was *Vesper*. The Morning Star was named *Phosphoros* ("light-bringer"), for when the Morning Star appeared the Sun and its light were not far behind. (The chemical element "phosphorus" —Latin spelling—was so named because it glowed in the dark as the result of slow combination with oxygen.) The Latin name for the Morning Star was *Lucifer*. Which also means "light-bringer."

Now notice that the Greeks made no use of mythology here. Their words for the Evening Star and Morning Star were logical, descriptive words. But then (during the sixth century B.C.) the Greek scholar, Pythagoras of Samos, arrived back in the Greek world after his travels in Babylonia. He brought with him a skullfull of Babylonian notions.

At the time, Babylonian astronomy was well developed and far in advance of the Greek bare beginnings. The Babylonian interest in astronomy was chiefly astrological in nature and so it seemed natural for them to equate the powerful planets with the powerful gods. (Since both had power over human beings, why not?) The Babylonians knew that the Evening Star and the Morning Star were a single planet—after all, they never appeared on the same day; if one was present, the other was absent, and it was clear from their movements that the Morning Star passed the Sun and became the Evening Star and vice versa. Since the planet representing both was so bright and beautiful, the Babylonians very logically felt it appropriate to equate it with Ishtar, their goddess of beauty and love.

Pythagoras brought back to Greece this Babylonian knowledge of the oneness of the Evening and Morning Star, and Hesperos and Phosphoros vanished from the heavens. Instead, the Babylonian system was copied and the planet was named for the Greek goddess of beauty and love, Aphrodite. To the Romans this was their corresponding goddess *Venus*, and so it is to us.

Thus, the habit of naming heavenly bodies for gods and goddesses was, it seems to me, deliberately copied from the Babylonians (and their predecessors) by the Greeks.

The name "Venus," by the way, represents a problem. Adjectives from these classical words have to be taken from the genitive case and the genitive form of "Venus" is

Veneris. (Hence, "venerable" for anything worth the respect paid by the Romans to the goddess; and because the Romans respected old age, "venerable" came to be applied to old men rather than young women.)

So we cannot speak of "Venusian atmosphere" or "Venutian atmosphere" as science fiction writers sometimes do. We must say "Venerian atmosphere." Unfortunately, this has uncomfortable associations and it is not used. We might turn back to the Greek name but the genitive form there is *Aphrodisiakos,* and if we speak of the "Aphrodisiac atmosphere" I think we will give a false impression.

But something must be done. We are actually exploring the atmosphere of Venus with space probes and some adjective is needed. Fortunately, there is a way out. The Venus cult was very prominent in early days in a small island south of Greece. It was called Kythera (Cythera in Latin spelling) so that Aphrodite was referred to, poetically, as the "Cytherean goddess." Our poetic astronomers have therefore taken to speaking of the "Cytherean atmosphere."

The other four planets present no problem. The second brightest planet is truly the king planet. Venus may be brighter but it is confined to the near neighborhood of the Sun and is never seen at midnight. The second brightest, however, can shine through all the hours of night and so it should fittingly be named for the chief god. The Babylonians accordingly named it "Marduk." The Greeks followed suit and called it "Zeus," and the Romans named it *Jupiter*. The genitive form of Jupiter is *Jovis*, so that we speak of the "Jovian satellites." A person born under the astrological influence of Jupiter is "jovial."

Then there is a reddish planet and red is obviously the color of blood; that is, of war and conflict. The Babylonians named this planet "Nergal" after their god of war, and the Greeks again followed suit by naming it "Ares" after theirs. Astronomers who study the surface features of the planet are therefore studying "areography." The Latins used their god of war, *Mars,* for the planet. The genitive form is *Martis*, so we can speak of the "Martian canals."

The planet nearest the Sun, appears, like Venus, as both an evening star and morning star. Being smaller and less reflective than Venus, as well as closer to the Sun, it is much harder to see. By the time the Greeks got around to naming it, the mythological notion had taken hold. The evening star manifestation was named "Hermes," and the morning star one "Apollo."

The latter name is obvious enough, since the later Greeks associated Apollo with the Sun, and by the time the planet Apollo was in the sky the Sun was due very shortly. Because the planet was closer to the Sun than any other planet (though, of course, the Greeks did not know this was the reason), it moved more quickly against the stars than any object but the Moon. This made it resemble the wing-footed messenger of the gods, Hermes. But giving the planet two names was a matter of conservatism. With the Venus matter straightened out, Hermes/Apollo was quickly reduced to a single planet and Apollo was dropped. The Romans named it "Mercurius," which was their equivalent of Hermes, and we call it *Mercury*. The quick journey of Mercury across the stars is like the lively behavior of droplets of quicksilver, which came to be called "mercury," too, and we know the type of personality that is described as "mercurial."

There is one planet left. This is the most slowly moving of all the planets known to the ancient Greeks (being the farthest from the Sun) and so they gave it the name of an ancient god, one who would be expected to move in grave and solemn steps. They called it "Cronos," the father of Zeus and ruler of the universe before the successful revolt of the Olympians under Zeus's leadership. The Romans gave it the name of a god they considered the equivalent of Cronos and called it "Saturnus," which to us is *Saturn.* People born under Saturn are supposed to reflect its gravity and are "saturnine."

For two thousand years the Earth, Sun, Moon, Mercury, Venus, Mars, Jupiter, and Saturn remained the only known bodies of the Solar System. Then came 1610 and the Italian astronomer Galileo Galilei, who built himself a telescope and turned it on the heavens. In no time at all he found four subsidiary objects circling the planet Jupiter. (The German astronomer Johann Kepler promptly named such subsidiary bodies "satellites," from a Latin word for the hangers-on of some powerful man.)

There was a question as to what to name the new bodies. The mythological names of the planets had hung on into the Christian era, but I imagine there must have been some natural hesitation about using heathen gods for new bodies. Galileo himself felt it wise to honor Cosimo Medici II, Grand Duke of Tuscany from whom he expected (and later received) a position, and called them *Sidera Medicea* (the Medicean stars). Fortunately this didn't stick. Nowadays we call the four satellites the "Galilean satellites" as a group, but individually we use mythological names after all. A German astronomer, Simon Marius, gave them these names after having discovered the satellites one day later than Galileo.

The names are all in honor of Jupiter's (Zeus's) loves, of which there were many. Working outward from Jupiter, the first is *Io* (two syllables please, eye′oh), a maiden whom Zeus turned into a heifer to hide her from his wife's jealousy. The second is *Europa,* whom Zeus in the form of a bull abducted from the coast of Phoenicia in Asia and carried to Crete (which is how Europe received its name). The third is *Ganymede,* a young Trojan lad (well, the Greeks were liberal about such things) whom Zeus abducted by assuming the guise of an eagle. And the fourth is *Callisto,* a nymph whom Zeus's wife caught and turned into a bear.

As it happens, naming the third satellite for a male rather than for a female turned out to be appropriate, for Ganymede is the largest of the Galilean satellites and, indeed, is the largest of any satellite in the Solar System. (It is even larger than Mercury, the smallest planet.)

The naming of the Galilean satellites established once and for all the convention that bodies of the Solar System were to be named mythologically, and except in highly unusual instances this custom has been followed since.

In 1655 the Dutch astronomer Christian Huygens discovered a satellite of Saturn (now known to be the sixth from the planet). He named it *Titan*. In a way this was appropriate, for Saturn (Cronos) and his brothers and sisters, who ruled the Universe before Zeus took over, were referred to collectively as "Titans." However, since the name refers to a group of beings and not to an individual being, its use is unfortunate. The name was appropriate in a second fashion, too. "Titan" has come to mean "giant" because the Titans and their allies were pictured by the Greeks as being of superhuman size (whence the word "titanic"), and it turned out that Titan was one of the largest satellites in the Solar System.

The Italian-French astronomer Gian Domenico Cassini was a little more precise than Huygens had been. Between 1671 and 1684 he discovered four more satellites of Saturn, and these he named after individual Titans and Titanesses. The satellites now known to be 3rd, 4th, and 5th from Saturn he named *Tethys, Dione,* and *Rhea,* after three sisters of Saturn. Rhea was Saturn's wife as well. The 8th satellite from Saturn he named *Iapetus* after one of Saturn's brothers. (Iapetus is frequently mispronounced. In English it is "eye-ap'ih-tus.") Here finally the Greek names were used, chiefly because there were no Latin equivalents, except for Rhea. There the Latin equivalent is *Ops*. Cassini tried to lump the four satellites he had discovered under the name of "Ludovici" after his patron, Louis XIV—*Ludovicus*, in Latin—but that second attempt to honor royalty also failed.

And so within 75 years after the discovery of the telescope, nine new bodies of the Solar System were discovered, four satellites of Jupiter and five of Saturn. Then something more exciting turned up.

On March 13, 1781, a German-English astronomer, William Herschel, surveying the heavens, found what he thought was a comet. This, however, proved quickly to be no comet at all, but a new planet with an orbit outside that of Saturn.

There arose a serious problem as to what to name the

new planet, the first to be discovered in historic times. Herschel himself called it "Georgium Sidus" ("George's star") after his patron, George III of England, but this third attempt to honor royalty failed. Many astronomers felt it should be named for the discoverer and called it "Herschel." Mythology, however, won out.

The German astronomer Johann Bode came up with a truly classical suggestion. He felt the planets ought to make a heavenly family. The three innermost planets (excluding the Earth) were Mercury, Venus, and Mars, who were siblings, and children of Jupiter, whose orbit lay outside theirs. Jupiter in turn was the son of Saturn, whose orbit lay outside his. Since the new planet had an orbit outside Saturn's, why not name it for *Uranus*, god of the sky and father of Saturn? The suggestion was accepted and Uranus* it was. What's more, in 1798 a German chemist, Martin Heinrich Klaproth, discovered a new element he named in its honor as "uranium."

In 1787 Herschel went on to discover Uranus's two largest satellites (the 4th and 5th from the planet, we now know). He named them from mythology, but *not* from Graeco-Roman mythology. Perhaps, as a naturalized Englishman, he felt 200 per cent English (it's that way, sometimes) so he turned to English folktales and named the satellites *Titania* and *Oberon*, after the queen and king of the fairies (who make an appearance, notably, in Shakespeare's *A Midsummer Night's Dream*).

In 1789 he went on to discover two more satellites of Saturn (the two closest to the planet) and here too he disrupted mythological logic. The planet and the five satellites then known were all named for various Titans and Titanesses (plus the collective name, Titan). Herschel named his two *Mimas* and *Enceladus* (en-sel′a-dus) after two of the giants who rose in rebellion against Zeus long after the defeat of the Titans.

After the discovery of Uranus, astronomers climbed hungrily upon the discover-a-planet bandwagon and searched

* Uranus is pronounced "yoo′ruh-nus." I spent almost all my life accenting the second syllable and no one ever corrected me. I just happened to be reading Webster's Unabridged one day . . .

particularly in the unusually large gap between Mars and Jupiter. The first to find a body there was the Italian astronomer Giuseppe Piazzi. From his observatory at Palermo, Sicily he made his first sighting on January 1, 1801.

Although a priest, he adhered to the mythological convention and named the new body *Ceres,* after the tutelary goddess of his native Sicily. She was a sister of Jupiter and the goddess of grain (hence "cereal") and agriculture. This was the second planet to receive a feminine name (Venus was the first, of course) and it set a fashion. Ceres turned out to be a small body (485 miles in diameter), and many more were found in the gap between Mars and Jupiter. For a hundred years, all the bodies so discovered were given feminine names.

Three "planetoids" were discovered in addition to Ceres over the next six years. Two were named *Juno* and *Vesta* after Ceres' two sisters. They were also the sisters of Jupiter, of course, and Juno was his wife as well. The remaining planetoid was named *Pallas*, one of the alternate names for Athena, daughter of Zeus (Jupiter) and therefore a niece of Ceres. (Two chemical elements discovered in that decade were named "cerium" and "palladium" after Ceres and Pallas.)

Later planetoids were named after a variety of minor goddesses, such as *Hebe*, the cupbearer of the gods, *Iris,* their messenger, the various Muses, Graces, Horae, nymphs, and so on. Eventually the list was pretty well exhausted and planetoids began to receive trivial and foolish names. We won't bother with those.

New excitement came in 1846. The motions of Uranus were slightly erratic, and from them the Frenchman Urbain J. J. Leverrier and the Englishman John Couch Adams calculated the position of a planet beyond Uranus, the gravitational attraction of which would account for Uranus's anomalous motion. The planet was discovered in that position.

Once again there was difficulty in the naming. Bode's mythological family concept could not be carried on, for

Uranus was the first god to come out of chaos and had no father. Some suggested the planet be named for Leverrier. Wiser council prevailed. The new planet, rather greenish in its appearance, was named *Neptune* after the god of the sea.

(Leverrier also calculated the possible existence of a planet inside the orbit of Mercury and named it *Vulcan,* after the god of fire and the forge, a natural reference to the planet's closeness to the central fire of the Solar System. However, such a planet was never discovered and undoubtedly does not exist.)

As soon as Neptune was discovered, the English astronomer William Lassell turned his telescope upon it and discovered a large satellite which he named *Triton,* appropriately enough, since Triton was a demigod of the sea and a son of Neptune (Poseidon).

In 1851 Lassell discovered two more satellites of Uranus, closer to the planet than Herschel's Oberon and Titania. Lassell, also English, decided to continue Herschel's English folklore bit. He turned to Alexander Pope's *The Rape of the Lock,* wherein were two elfish characters, *Ariel* and *Umbriel,* and these names were given to the satellites.

More satellites were turning up. Saturn was already known to have seven satellites, and in 1848 the American astronomer George P. Bond discovered an eigth; in 1898 the American astronomer William H. Pickering discovered a ninth and completed the list. These were named *Hyperion* and *Phoebe* after a Titan and Titaness. Pickering also thought he had discovered a tenth in 1905, and named it *Themis,* after another Titaness, but this proved to be mistaken.

In 1877 the American astronomer Asaph Hall, waiting for an unusually close approach of Mars, studied its surroundings carefully and discovered two tiny satellites, which he named *Phobos* ("fear") and *Deimos* ("terror"), two sons of Mars (Ares) in Greek legend, though obviously mere personifications of the inevitable consequences of Mars's pastime of war.

In 1892 another American astronomer, Edward E. Barnard, discovered a fifth satellite of Jupiter, closer than the Galilean satellites. For a long time it received no name, being called "Jupiter V" (the fifth to be discovered) or "Barnard's satellite." Mythologically, however, it was given the name *Amalthea* by the French astronomer Camille Flammarion, and this is coming into more common use. I am glad of this. Amalthea was the nurse of Jupiter (Zeus) in his infancy, and it is pleasant to have the nurse of his childhood closer to him than the various girl and boy friends of his maturer years.

In the twentieth century no less than seven more Jovian satellites were discovered, all far out, all quite small, all probably captured planetoids, all nameless. Unofficial names have been proposed. Of these, the three planetoids nearest Jupiter bear the names *Hestia, Hera,* and *Demeter,* after the Greek names of the three sisters of Jupiter (Zeus). Hera, of course, is his wife as well. Under the Roman versions of the names (Vesta, Juno, and Ceres, respectively) all three are planetoids. The two farthest are *Poseidon* and *Hades,* the two brothers of Jupiter (Zeus). The Roman version of Poseidon's name (Neptune) is applied to a planet. Of the remaining satellites, one is *Pan,* a grandson of Jupiter (Zeus), and the other is *Adrastea,* another of the nurses of his infancy.

The name of Jupiter's (Zeus's) wife, Hera, is thus applied to a satellite much farther and smaller than those commemorating four of his extracurricular affairs. I'm not sure that this is right, but I imagine astronomers understand these things better than I do.

In 1898 the German astronomer G. Witt discovered an unusual planetoid, one with an orbit that lay closer to the Sun than did any other of the then-known planetoids. It inched past Mars and came rather close to Earth's orbit. Not counting the Earth, this planetoid might be viewed as passing between Mars and Venus and therefore Witt gave it the name of *Eros*, the god of love, and the son of Mars (Ares) and Venus (Aphrodite).

This started a new convention, that of giving planetoids

with odd orbits masculine names. For instance, the planetoids that circle in Jupiter's orbit all received the names of masculine participants in the Trojan war: *Achilles, Hector, Patroclus, Ajax, Diomedes, Agamemnon, Priamus, Nestor, Odysseus, Antilochus, Aeneas, Anchises,* and *Troilus.*

A particularly interesting case arose in 1948, when the German-American astronomer Walter Baade discovered a planetoid that penetrated more closely to the Sun than even Mercury did. He named it *Icarus,* after the mythical character who flew too close to the Sun, so that the wax holding the feathers of his artificial wings melted, with the result that he fell to his death.

Two last satellites were discovered. In 1948 a Dutch-American astronomer, Gerard P. Kuiper, discovered an innermost satellite of Uranus. Since Ariel (the next innermost) is a character in William Shakespeare's *The Tempest* as well as in Pope's *The Rape of the Lock,* free association led Kuiper to the heroine of *The Tempest* and he named the new satellite *Miranda.*

In 1950 he discovered a second satellite of Neptune. The first satellite, Triton, represents not only the name of a particular demigod, but of a whole class of merman-like demigods of the sea. Kuiper named the second, then, after a whole class of mermaid-like nymphs of the sea, *Nereid.*

Meanwhile, during the first decades of the twentieth century, the American astronomer Percival Lowell was searching for a ninth planet beyond Neptune. He died in 1916 without having succeeded but in 1930, from his observatory and in his spirit, Clyde W. Tombaugh made the discovery.

The new planet was named *Pluto*, after the god of the Underworld, as was appropriate since it was the planet farthest removed from the light of the Sun. (And in 1940, when two elements were found beyond uranium, they were named "neptunium" and "plutonium," after Neptune and Pluto, the two planets beyond Uranus.)

Notice, though, that the first two letters of "Pluto" are the initials of Percival Lowell. And so, finally, as astronomer got his name attached to a planet. Where Herschel and Leverrier had failed, Percival Lowell had succeeded, at least by initial, and under cover of the mythological conventions.

6. ROUND AND ROUND AND . . .

Anyone who writes a book on astronomy for the general public eventually comes up against the problem of trying to explain that the Moon always presents one face to the earth, but is nevertheless rotating.

To the average reader who has not come up against this problem before and who is impatient with involved subtleties, this is a clear contradiction in terms. It is easy to accept the fact that the Moon always presents one face to the Earth because even to the naked eye, the shadowy blotches on the Moon's surface are always found in the same position. But in that case it seems clear that the Moon is not rotating, for if it were rotating we would, bit by bit, see every portion of its surface.

Now it is no use smiling gently at the lack of sophistication of the average reader, because he happens to be right. The Moon is *not* rotating with respect to the observer on the Earth's surface. When the astronomer says that the Moon *is* rotating, he means with respect to other observers altogether.

For instance, if one watches the Moon over a period of time, one can see that the line marking off the sunlight from the shadow progresses steadily around the Moon; the Sun shines on every portion of the Moon in steady progression. This means that to an observer on the surface of the Sun (and there are very few of those), the Moon would seem to be rotating, for the observer would, little by little, see every portion of the Moon's surface as it turned to be exposed to the sunlight.

But our average reader may reason to himself as follows: "I see only one face of the Moon and I say it is *not* rotating. An observer on the Sun sees all parts of the Moon and he says it *is* rotating. Clearly, I am more important than the Sun observer since, firstly, I exist and he doesn't, and secondly, even if he existed, I am me and he isn't. Therefore, I insist on having it my way. The Moon does *not* rotate!"

There has to be a way out of this confusion, so let's think things through a little more systematically. And to do so, let's start with the rotation of the Earth itself, since that is a point nearer to all our hearts.

One thing we can admit to begin with: To an observer on the Earth, the Earth is not rotating. If you stay in one place from now till doomsday, you will see but one portion of the Earth's surface and no other. As far as you are concerned, the planet is standing still. Indeed, through most of civilized human history, even the wisest of men insisted that "reality" (whatever that may be) exactly matched the appearance and that the Earth "really" did not rotate. As late as 1633, Galileo found himself in a spot of trouble for maintaining otherwise.

But suppose we had an observer on a star situated (for simplicity's sake) in the plane of the Earth's equator; or, to put it another way, on the celestial equator (see Chapter 3). Let us further suppose that the observer was equipped with a device that made it possible for him to study the Earth's surface in detail. To him, it would seem that the Earth rotated, for little by little he would see every part of its surface pass before his eyes. By taking note of some particular small feature (for example, you and I standing on some point on the equator) and timing its return, he could even determine the exact period of the Earth's rotation—that is, as far as he is concerned.

We can duplicate his feat, for when the observer on the star sees us exactly in the center of that part of Earth's surface visible to himself, we in turn see the observer's star directly overhead. And just as he would time the periodic return of ourselves to that centrally located position, so we could time the return of his star to the overhead point. The period determined will be the same in either case. (Let's measure this time in minutes, by the way. A minute can be defined as 60 seconds, where 1 second is equal to 1/31,556,925.9747 of the tropical year.)

The period of Earth's rotation with respect to the star is just about 1436 minutes. It doesn't matter which star we use, for the apparent motion of the stars with respect to one another, as viewed from the Earth, is so vanishingly small that the constellations can be considered as moving all in one piece.

The period of 1436 minutes is called Earth's "sidereal day." The word "sidereal" comes from a Latin word for "star," and the phrase therefore means, roughly speaking, "the star-based day."

Suppose, though, that we were considering an observer on the Sun. If he were watching the Earth, he, too, would observe it rotating, but the period of rotation would not seem the same to him as to the observer on the star. Our solar observer would be much closer to the Earth; close enough, in fact, for Earth's motion about the Sun to introduce a new factor. In the course of a single rotation of the Earth (judging by the star's observer), the Earth would have moved an appreciable distance through space, and the solar observer would find himself viewing the planet from a different angle. The Earth would have to turn for four more minutes before it adjusted itself to the new angle of view.

We could interpret these results from the point of view of an observer on the Earth. To duplicate the measurements of the solar observer, we on Earth would have to measure the period of time from one passage of the Sun overhead to the next (from noon to noon, in other words). Because of the revolution of the Earth about the Sun, the Sun seems to move from west to east against the background of the stars. After the passage of one sidereal day, a particular star would have returned to the overhead position, but the Sun would have drifted eastward to a point where four more minutes would be required to make it pass overhead. The solar day is therefore 1440 minutes long, 4 minutes longer than the sidereal day.

Next, suppose we have an observer on the Moon. He is even closer to the Earth and the apparent motion of the Earth against the stars is some thirteen times greater for him than for an observer on the Sun. Therefore, the discrepancy between what he sees and what the star observer sees is about thirteen times greater than is the Sun/star discrepancy.

If we consider this same situation from the Earth, we would be measuring the time between successive passages of the Moon exactly overhead. The Moon drifts eastward against the starry background at thirteen times the rate the Sun does. After one sidereal day is completed, we have to wait a total of 54 additional minutes for the Moon to pass overhead again. The Earth's "lunar day" is therefore 1490 minutes long.

We could also figure out the periods of Earth's rotation with respect to an observer on Venus, on Jupiter, on Halley's Comet, on an artificial satellite, and so on, but I shall have mercy and refrain. We can instead summarize the little we do have as follows:

sidereal day	1436 minutes
solar day	1440 minutes
lunar day	1490 minutes

By now it may seem reasonable to ask: But which is *the* day? The *real* day?

The answer to that question is that the question is not a reasonable one at all, but quite unreasonable; and that there is no real day, no real period of rotation. There are only different *apparent* periods, the lengths of which depend upon the position of the observer. To use a prettier-sounding phrase, the length of the period of the Earth's rotation depends on the frame of reference, and all frames of reference are equally valid.

But if all frames of reference are equally valid, are we left nowhere?

Not at all! Frames of reference may be equally valid, but they are usually not equally useful. In one respect, a particular frame of reference may be most useful; in another respect, another frame of reference may be most useful. We are free to pick and choose, using now one, now another, exactly as suits our dear little hearts.

For instance, I said that the solar day is 1440 minutes long but actually that's a lie. Because the Earth's axis is tipped to the plane of its orbit and because the Earth is sometimes closer to the Sun and sometimes farther (so that it moves now faster, now slower in its orbit), the solar day is sometimes a little longer than 1440 minutes and sometimes a little shorter. If you mark off "noons" that are exactly 1440 minutes apart all through the year, there will be times during the year when the Sun will pass overhead fully 16 minutes ahead of schedule, and other times when it will pass overhead fully 16 minutes behind schedule. Fortunately, the errors cancel out and by the end of the year all is even again.

For that reason it is not the solar day itself that is 1440 minutes long, but the average length of all the solar days of the year; this average is the "mean solar day." And at noon of all but four days a year, it is not the real Sun that crosses the overhead point but a fictitious body called the "mean Sun." The mean Sun is located where the real Sun would be if the real Sun moved perfectly evenly.

The lunar day is even more uneven than the solar day, but the sidereal day is a really steady affair. A particular star passes overhead every 1436 minutes virtually on the dot.

If we're going to measure time, then, it seems obvious that the sidereal day is the most useful, since it is the most constant. Where the sidereal day is used as the basis for checking the clocks of the world by the passage of a star across the hairline of a telescope eyepiece, then the Earth itself, as it rotates with respect to the stars, is serving as the reference clock. The second can then be defined as 1/1436.09 of a sidereal day. (Actually, the length of the year is even more constant than that of the sidereal day, which is why the second is now officially defined as a fraction of the tropical year.)

The solar day, uneven as it is, carries one important advantage. It is based on the position of the Sun, and the position of the Sun determines whether a particular portion of the Earth is in light or in shadow. In short, the solar day is equal to one period of light (daytime) plus one period of darkness (night). The average man throughout history has managed to remain unmoved by the position of the stars, and couldn't have cared less when one of them moved overhead; but he certainly couldn't help noticing, and even being deeply concerned, by the fact that it might be day or night at a particular moment; sunrise or sunset; noon or twilight.

It is the solar day, therefore, which is by far the most useful and important day of all. It was the original basis of time measurement and it is divided into exactly 24 hours, each of which is divided into 60 minutes (and 24 times 60 is 1440, the number of minutes in a solar day). On this basis, the sidereal day is 23 hours 56 minutes long and the lunar day is 24 hours 50 minutes long.

So useful is the solar day, in fact, that mankind has become accustomed to thinking of it as *the* day, and of thinking that the Earth "really" rotates in exactly 24 hours, where actually this is only its apparent rotation with respect to the Sun, no more "real" or "unreal" than its apparent rotation with respect to any other body. It is no more "real" or "unreal," in fact, than the apparent rotation of the Earth with respect to an observer on the Earth—that is, to the apparent lack of rotation altogether.

The lunar day has its uses, too. If we adjusted our watches to lose 2 minutes 5 seconds every hour, it would then be running on a lunar day basis. In that case, we would find that high tide (or low tide) came exactly twice a day and at the same times every day—indeed, at twelve-hour intervals (with minor variations).

And extremely useful is the frame of reference of the Earth itself; to wit, the assumption that the Earth is not rotating at all. In judging a billiard shot, in throwing a baseball, in planning a trip cross-country, we never take into account any rotation of the Earth. We always assume the Earth is standing still.

Now we can pass on to the Moon. For the viewer from the Earth, as I said earlier, it does not rotate at all so that its "terrestrial day" is of infinite length. Nevertheless, we can maintain that the Moon rotates if we shift our frame of reference (usually without warning or explanation so that the reader has trouble following). As a matter of fact, we can shift our plane of reference to either the Sun or the stars so that not only can the Moon be considered to rotate but to do so in either of two periods.

With respect to the stars, the period of the Moon's rotation is 27 days, 7 hours, 43 minutes, 11.5 seconds, or 27.3217 days (where the day referred to is the 24-hour mean solar day). This is the Moon's sidereal day. It is also the period (with respect to the stars) of its revolution about the Earth, so it is almost invariably called the "sidereal month."

In one sidereal month, the Moon moves about $\frac{1}{13}$ of the length of its orbit about the Sun, and to an observer on the Sun the change in angle of viewpoint is considerable. The Moon must rotate for over two more days to make up for it. The period of rotation of the Moon with respect to the Sun is the same as its period of revolution about the Earth with respect to the Sun, and this may be called the Moon's solar day or, better still, the solar month. (As a matter of fact, as I shall shortly point out, it is called neither.) The solar month is 29 days, 12 hours, 44 minutes, 2.8 seconds long, or 29.5306 days long.

Of these two months, the solar month is far more useful to mankind because the phases of the Moon depend on the relative positions of the Moon and Sun. It is therefore 29.5306 days, or one solar month, from new Moon to new Moon, or from full Moon to full Moon. In ancient times, when the phases of the Moon were used to mark off the seasons, the solar month became the most important unit of time.

Indeed, great pains were taken to detect the exact day on which successive new Moons appeared in order that the calendar be accurately kept (see Chapter 1). It was the place of the priestly caste to take care of this, and the very word "calendar," for instance, comes from the Latin word meaning "to proclaim," because the beginning of each month was proclaimed with much ceremony. An assembly of priestly officials, such as those that, in ancient times, might have proclaimed the beginning of each month, is called a "synod." Consequently, what I have been calling the solar month (the logical name) is, actually, called the "synodic month."

The farther a planet is from the Sun and the faster it turns with respect to the stars, the smaller the discrepancy between its sidereal day and solar day. For the planets beyond Earth, the discrepancy can be ignored.

For the two planets closer to the Sun than the Earth the discrepancy is very great. Both Mercury and Venus turn one face eternally to the Sun and have no solar day. They turn with respect to the stars, however, and have a sidereal day which turns out to be as long as the period of their revolution about the Sun (again with respect to the stars).

If the various true satellites of the Solar System (see Chapter 7) keep one face to their primaries at all times, as is very likely true, their sidereal day would be equal to their period of revolution about their primary.

If this is so I can prepare a table (not quite like any I have ever seen) listing the sidereal period of rotation for each of the 32 major bodies of the Solar System: the Sun, the Earth, the eight other planets (even Pluto, which has a rotation figure, albeit an uncertain one), the Moon, and the 21 other true satellites. For the sake of direct comparison I'll give the period in minutes and list them in the order of length. After each satellite I shall put the name of the primary in parentheses and give a number to represent the position of that satellite, counting outward from the primary.

These figures represent the time it takes for stars to make a complete circuit of the skies from the frame of reference of an observer on the surface of the body in question. If you divide each figure by 720, you get the number of minutes it would take a star (in the region of the body's celestial equator) to travel the width of the Sun or Moon as seen from the Earth.

Body	*Sidereal Day (minutes)*
Venus	324,000
Mercury	129,000
Iapetus (Saturn-8)	104,000
Moon (Earth-1)	39,300
Sun	35,060
Hyperion (Saturn-7)	30,600
Callisto (Jupiter-5)	24,000
Titan (Saturn-6)	23,000
Oberon (Uranus-5)	19,400
Titania (Uranus-4)	12,550
Ganymede (Jupiter-4)	10,300
Pluto	8650
Triton (Neptune-1)	8450
Rhea (Saturn-5)	6500
Umbriel (Uranus-3)	5950
Europa (Jupiter-3)	5100
Dione (Saturn-4)	3950
Ariel (Uranus-2)	3630
Tethys (Saturn-3)	2720
Io (Jupiter-2)	2550
Miranda (Uranus-1)	2030
Enceladus (Saturn-2)	1975
Deimos (Mars-2)	1815
Mars	1477
Earth	1436
Mimas (Saturn-1)	1350
Neptune	948
Amaltheia (Jupiter-1)	720
Uranus	645
Saturn	614
Jupiter	590
Phobos	460

On Earth itself, this takes about 2 minutes and no more, believe it or not. On Phobos (Mars's inner satellite), it takes only a little over half a minute. The stars will be whirling by at four times their customary rate, while a bloated Mars hangs motionless in the sky. What a sight that would be to see.

On the Moon, on the other hand, it would take 55 minutes for a star to cover the apparent width of the Sun. Heavenly bodies could be studied over continuous sustained intervals nearly thirty times as long as is possible on the Earth. I have never seen this mentioned as an advantage for a Moon-based telescope, but, combined with the absence of clouds or other atmospheric interference, it makes a lunar observatory something for which astronomers ought to be willing to undergo rocket trips.

On Venus, it would take 450 minutes or 7½ hours for a star to travel the apparent width of the Sun as we see it. What a fix astronomers could get on the heavens there—if only there were no clouds.

7. JUST MOONING AROUND

Almost every book on astronomy I have ever seen, large or small, contains a little table of the Solar System. For each planet, there's given its diameter, its distance from the sun, its time of rotation, its albedo, its density, the number of its moons, and so on.

Since I am morbidly fascinated by numbers, I jump on such tables with the perennial hope of finding new items of information. Occasionally, I am rewarded with such things as surface temperature or orbital velocity, but I never really get enough.

So every once in a while, when the ingenuity-circuits in my brain are purring along with reasonable smoothness, I deduce new types of data for myself out of the material on hand, and while away some idle hours. (At least I did this in the long-gone days when I had idle hours.)

I can still do it, however, provided I put the results into formal essay-form; so come join me and we will just moon around together in this fashion, and see what turns up.

Let's begin this way, for instance . . .

According to Newton, every object in the universe attracts every other object in the universe with a force (f) that is proportional to the product of the masses (m_1 and m_2) of the two objects divided by the square of the distance (d) between them, center to center. We multiply by the gravitational constant (g) to convert the proportionality to an equality, and we have:

$$f=\frac{gm_1m_2}{d^2} \qquad \text{(Equation 1)}$$

This means, for instance, that there is an attraction between the Earth and the Sun, and also between the Earth and the Moon, and between the Earth and each of the various planets and, for that matter, between the Earth and any meteorite or piece of cosmic dust in the heavens.

Fortunately, the Sun is so overwhelmingly massive compared with everything else in the Solar System that in calculating the orbit of the Earth, or of any other planet, an excellent first approximation is attained if only the planet and the Sun are considered, as though they were alone in the Universe. The effect of other bodies can be calculated later for relatively minor refinements.

In the same way, the orbit of a satellite can be worked out first by supposing that it is alone in the Universe with its primary.

It is at this point that something interests me. If the Sun is so much more massive than any planet, shouldn't it exert a considerable attraction on the satellite even though it is at a much greater distance from that satellite than the primary is? If so, just how considerable is "considerable"?

To put it another way, suppose we picture a tug of war going on for each satellite, with its planet on one side of the gravitational rope and the Sun on the other. In this tug of war, how well is the Sun doing?

I suppose astronomers have calculated such things, but I have never seen the results reported in any astronomy text, or the subject even discussed, so I'll do it for myself.

Here's how we can go about it. Let us call the mass of a satellite m, the mass of its primary (by which, by the way, I mean the planet in circles) m_p, and the mass of the Sun m_s. The distance from the satellite to its primary will be d_p, and the distance from the satellite to the sun will be d_s. The gravitational force between the satellite and its primary would be f_p and that between the satellite and the Sun would be f_s—and that's the whole business. I promise to use no other symbols in this chapter.

From Equation 1, we can say that the force of attraction between a satellite and its primary would be:

$$f_p = \frac{gmm_p}{d_p^2} \qquad \text{(Equation 2)}$$

while that between the same satellite and the Sun would be:

$$f_s = \frac{gmm_s}{d_s^2} \qquad \text{(Equation 3)}$$

What we are interested in is how the gravitational force between satellite and primary compares with that between satellite and Sun. In other words we want the ratio f_p/f_s, which we can call the "tug-of-war value." To get that we must divide equation 2 by equation 3. The result of such a division would be:

$$f_p/f_s = (m_p/m_s)\ (d_s/d_p)^2 \qquad \text{(Equation 4)}$$

In making the division, a number of simplifications have taken place. For one thing the gravitational constant has dropped out, which means we won't have to bother with an inconveniently small number and some inconvenient units. For another, the mass of the satellite has dropped out. (In other words, in obtaining the tug-of-war value, it doesn't matter how big or little a particular satellite is. The result would be the same in any case.)

What we need for the tug-of-war value (f_p/f_s), is the ratio of the mass of the planet to that of the sun (m_p/m_s) and the square of the ratio of the distance from satellite to Sun to the distance from satellite to primary $(d_s/d_p)^2$.

There are only six planets that have satellites and these, in order of decreasing distance from the Sun, are: Neptune, Uranus, Saturn, Jupiter, Mars, and Earth. (I place Earth at the end, instead of at the beginning, as natural chauvinism would dictate, for my own reasons. You'll find out.)

For these, we will first calculate the mass-ratio and the results turn out as follows:

Neptune	0.000052
Uranus	0.000044
Saturn	0.00028
Jupiter	0.00095
Mars	0.00000033
Earth	0.0000030

As you see, the mass ratio is really heavily in favor of the Sun. Even Jupiter, which is by far the most massive planet, is not quite one-thousandth as massive as the Sun. In fact, all the planets together (plus satellites, planetoids, comets, and meteoric matter) make up no more than 1/750 of the mass of the Sun.

So far, then, the tug of war is all on the side of the Sun. However, we must next get the distance ratio, and that favors the planet heavily, for each satellite is, of course, far closer to its primary than it is to the Sun. And what's more, this favorable (for the planet) ratio must be squared, making it even more favorable, so that in the end we can be reasonably sure that the Sun will lose out in the tug of war. But we'll check, anyway.

Let's take Neptune first. It has two satellites, Triton and Nereid. The average distance of each of these from the Sun is, of necessity, precisely the same as the average distance of Neptune from the Sun, which is 2,797,000,000 miles. The average distance of Triton from Neptune is, however, only 220,000 miles, while the average distance of Nereid from Neptune is 3,460,000 miles.

If we divide the distance from the Sun by the distance from Neptune for each satellite and square the result we get 162,000,000 for Triton and 655,000 for Nereid. We multiply each of these figures by the mass-ratio of Neptune to the Sun, and that gives us the tug-of-war value, which is:

Triton	8400
Nereid	34

The conditions differ markedly for the two satellites. The gravitational influence of Neptune on its nearer satellite, Triton, is overwhelmingly greater than the influence of the Sun on the same satellite. Triton is firmly in Neptune's grip. The outer satellite, Nereid, however, is attracted by Neptune considerably, but *not* overwhelmingly, more strongly than by the Sun. Furthermore, Nereid has a highly eccentric orbit, the most eccentric of any satellite in the system. It approaches to within 800,000 miles of Neptune at one end of its orbit and recedes to as far as 6 million miles at the other end. When most distant from Neptune, Nereid experiences a tug-of-war value as low as 11!

For a variety of reasons (the eccentricity of Nereid's orbit, for one thing) astronomers generally suppose that Nereid is not a true satellite of Neptune, but a planetoid captured by Neptune on the occasion of a too-close approach.

Neptune's weak hold on Nereid certainly seems to support this. In fact, from the long astronomic view, the association between Neptune and Nereid may be a temporary one. Perhaps the disturbing effect of the solar pull will eventually snatch it out of Neptune's grip. Triton, on the other hand, will never leave Neptune's company short of some catastrophe on a System-wide scale.

There's no point in going through all the details of the calculations for all the satellites. I'll do the work on my own and feed you the results. Uranus, for instance, has five known satellites, all revolving in the plane of Uranus's equator and all considered true satellites by astronomers. Reading outward from the planet, they are: Miranda, Ariel, Umbriel, Titania, and Oberon.

The tug-of-war values for these satellites are:

Miranda	24,600
Ariel	9850
Umbriel	4750
Titania	1750
Oberon	1050

All are safely and overwhelmingly in Uranus's grip, and the high tug-of-war values fit with their status as true satellites.

We pass on, then, to Saturn, then, to Saturn, which has nine satellites: Mimas, Enceladus, Tethys, Dione, Rhea, Titan, Hyperion, Iapetus, and Phoebe. Of these, the eight innermost revolve in the plane of Saturn's equator and are considered true satellites. Phoebe, the ninth, has a highly inclined orbit and is considered a captured planetoid.

The tug-of-war values for these satellites are:

Mimas	15,500
Enceladus	9800
Tethys	6400
Dione	4150
Rhea	2000
Titan	380
Hyperion	260
Iapetus	45
Phoebe	3½

Note the low value for Phoebe.

Jupiter has twelve satellites and I'll take them in two installments. The first five: Amaltheia, Io, Europa, Ganymede, and Callisto all revolve in the plane of Jupiter's equator and all are considered true satellites. The tug-of-war values for these are:

Amaltheia	18,200
Io	3260
Europa	1260
Ganymede	490
Callisto	160

and all are clearly in Jupiter's grip.

Jupiter, however, has seven more satellites which have no official names (see Chapter 5), and which are commonly known by Roman numerals (from VI to XII) given in the order of their discovery. In order of distance from Jupiter, they are VI, X, VII, XII, XI, VIII, and IX. All are small and with orbits that are eccentric and highly inclined to the plane of Jupiter's equator. Astronomers consider them captured planetoids. (Jupiter is far more massive than the other planets and is nearer the planetoid belt, so it is not surprising that it would capture seven planetoids.)

The tug-of-war results for these seven certainly bear out the captured planetoid notion, for the values are:

VI	4.4
X	4.3
VII	4.2
XII	1.3
XI	1.2
VIII	1.03
IX	1.03

Jupiter's grip on these outer satellites is feeble indeed.

Mars has two satellites, Phobos and Deimos, each tiny and very close to Mars. They rotate in the plane of Mars's equator, and are considered true satellites. The tug-of-war values are:

Phobos	195
Deimos	32

So far I have listed 30 satellites, of which 21 are considered true satellites and 9 are usually tabbed as (probably) captured planetoids. I would like, for the moment, to leave out of consideration the 31st satellite, which happens to be our own Moon (I'll get back to it, I promise) and summarize the 30 as follows:

	Number of Satellites	
Planet	*true*	*captured*
Neptune	1	1
Uranus	5	0
Saturn	8	1
Jupiter	5	7
Mars	2	0

It is unlikely that any additional true satellites will be discovered (though, to be sure, Miranda was discovered as recently as 1948). However, any number of tiny bodies coming under the classification of captured planetoids may yet turn up, particularly once we go out there and actually look.

But now let's analyze this list in terms of tug-of-war values. Among the true satellites the lowest tug-of-war value is that of Deimos, 32. On the other hand, among the nine satellites listed as captured, the highest tug-of-war value is that of Nereid with an average of 34.

Let us accept this state of affairs and assume that the tug-of-war figure 30 is a reasonable minimum for a true satellite and that any satellite with a lower figure is, in all likelihood, a captured and probably temporary member of the planet's family.

Knowing the mass of a planet and its distance from the Sun, we can calculate the distance from the planet's center at which this tug-of-war value will be found. We can use Equation 4 for the purpose, setting f_p/f_s equal to 30, putting in the known values for m_p, m_s, and d_s, and then solving for d_p. That will be the maximum distance at which we can expect to find a true satellite. The only planet that can't be handled in this way is Pluto, for which the value of m_p is very uncertain, but I omit Pluto cheerfully.

We can also set a minimum distance at which we can expect a true satellite; or, at least, a true satellite in the usual form. It has been calculated that if a true satellite is closer to its primary than a certain distance, tidal forces will break it up into fragments. Conversely, if fragments already exist at such a distance, they will not coalesce into a single body. This limit of distance is called the "Roche limit" and is named for the astronomer E. Roche, who worked it out in 1849. The Roche limit is a distance from a planetary center equal to 2.44 times the planet's radius.

So, sparing you the actual calculations, here are the results for the four outer planets:

Planet	*Distance of True Satellite (miles from the center of the primary)* maximum (tug-of-war = 30)	minimum (Roche limit)
Neptune	3,700,000	38,000
Uranus	2,200,000	39,000
Saturn	2,700,000	87,000
Jupiter	2,700,000	106,000

As you see, each of these outer planets, with huge masses and far distant from the competing Sun, has ample room for large and complicated satellite systems within these generous limits, and the 21 true satellites all fall within them.

Saturn does possess something within Roche's limit—its ring system. The outermost edge of the ring system stretches out to a distance of 85,000 miles from the planet's center. Obviously the material in the rings could have been collected into a true satellite if it had not been so near Saturn.

The ring system is unique as far as visible planets are concerned, but of course the only planets we can see are those of our own Solar System. Even of these, the only ones we can reasonably consider in connection with satellites (I'll explain why in a moment) are the four large ones.

Of these, Saturn has a ring system and Jupiter just barely misses one. Its innermost satellite, Amaltheia, is about 110,000 miles from the planet's center, with the Roche limit at 106,000 miles. A few thousand miles inward and Jupiter would have rings. I would like to make the suggestion therefore that once we reach outward to explore other stellar systems we will discover (probably to our initial amazement) that about half the large planets we find will be equipped with rings after the fashion of Saturn.

Next we can try to do the same thing for the inner planets. Since the inner planets are, one and all, much less massive than the outer ones and much closer to the competing Sun, we might guess that the range of distances open to true satellite formation would be more limited, and we would be right. Here are the actual figures as I have calculated them.

Thus, you see, where each of the outer planets has a range of two million miles or more within which true satellites could form, the situation is far more restricted for the inner planets. Mars and Venus have a permissible

Planet	*Distance of True Satellite (miles from the center of the primary)*	
	maximum (tug-of-war = 30)	*minimum (Roche limit)*
Mars	15,000	5150
Earth	29,000	9600
Venus	19,000	9200
Mercury	1300	3800

range of but 10,000 miles. Earth does a little better, with 20,000 miles.

Mercury is the most interesting case. The maximum distance at which it can expect to form a natural satellite against the overwhelming competition of the nearby Sun is well within the Roche limit. It follows from that, if my reasoning is correct, that Mercury *cannot* have a true satellite, and that anything more than a possible spattering of gravel is not to be expected.

In actual truth, no satellite has been located for Mercury but, as far as I know, nobody has endeavored to present a reason for this or treat it as anything other than an empirical fact. If any Gentle Reader, with a greater knowledge of astronomic detail than myself, will write to tell me that I have been anticipated in this, and by whom, I will try to take the news philosophically. At the very least, I will confine my kicking and screaming to the privacy of my study.

Venus, Earth, and Mars are better off than Mercury and do have a little room for true satellites beyond the Roche limit. It is not much room, however, and the chances of gathering enough material over so small a volume of space to make anything but a very tiny satellite is minute.

And, as it happens, neither Venus nor Earth has any satellite at all (barring possible minute chunks of gravel) within the indicated limits, and Mars has two small satellites, one perhaps 12 miles across and the other 6, which scarcely deserve the name.

It is amazing, and very gratifying to me, to note how all this makes such delightful sense, and how well I can reason out the details of the satellite systems of the various planets. It is such a shame that one small thing remains unaccounted for; one trifling thing I have ignored so far, but—

WHAT IN BLAZES IS OUR OWN MOON DOING WAY OUT THERE?

It's too far out to be a true satellite of the Earth, if we go by my beautiful chain of reasoning—which is too beautiful for me to abandon. It's too big to have been captured by the Earth. The chances of such a capture having been effected and the Moon then having taken up a nearly circular orbit about the Earth are too small to make such an eventuality credible.

There are theories, of course, to the effect that the Moon was once much closer to the Earth (within my permitted limits for a true satellite) and then gradually moved away as a result of tidal action. Well, I have an objection to that. If the Moon were a true satellite that originally had circled Earth at a distance of, say, 20,000 miles, it would almost certainly be orbiting in the plane of Earth's equator and it isn't.

But, then, if the Moon is neither a true satellite of the Earth nor a captured one, what is it? This may surprise you, but I have an answer; and to explain what that answer is, let's get back to my tug-of-war determinations. There is, after all, one satellite for which I have not calculated it, and that is our Moon. We'll do that now.

The average distance of the Moon from the Earth is 237,000 miles, and the average distance of the Moon from the Sun is 93,000,000 miles. The ratio of the Moon-Sun distance to the Moon-Earth distance is 392. Squaring that gives us 154,000. The ratio of the mass of the Earth to that of the Sun was given earlier in the chapter and is 0.0000030. Multiplying this figure by 154,000 gives us the tug-of-war value, which comes out to:

Moon	0.46

The Moon, in other words, is unique among the satellites of the Solar System in that its primary (us) *loses* the tug of war with the Sun. The Sun attracts the Moon twice as strongly as the Earth does.

We might look upon the Moon, then, as neither a true satellite of the Earth nor a captured one, but as a planet in its own right, moving about the Sun in careful step with the Earth. To be sure, from within the Earth-Moon system, the simplest way of picturing the situation is to have the Moon revolve about the Earth; but if you were to draw a picture of the orbits of the Earth and Moon about the Sun exactly to scale, you would see that the Moon's orbit is everywhere concave toward the Sun. It is always "falling" toward the Sun. All the other satellites, without exception, "fall" away from the Sun through part of their orbits, caught as they are by the superior pull of their primary—but not the Moon.

And consider this—the Moon does not revolve about the Earth in the plane of Earth's equator, as would be expected of a true satellite. Rather it revolves about the Earth in a plane quite close to that of the ecliptic; that is, to the plane in which the planets, generally, rotate about the Sun. This is just what would be expected of a planet!

Is it possible then, that there is an intermediate point between the situation of a massive planet far distant from the Sun, which develops about a single core, with numerous satellites formed, and that of a small planet near the Sun which develops about a single core with no satellites? Can there be a boundary condition, so to speak, in which there is condensation about two major cores so that a double planet is formed?

Maybe Earth just hit the edge of the permissible mass and distance; a little too small, a little too close. Perhaps if it were better situated the two halves of the double planet would have been more of a size. Perhaps both might have had atmospheres and oceans and—life. Perhaps in other stellar systems with a double planet, a greater equality is more usual.

What a shame if we have missed that . . .

Or, maybe (who knows), what luck!

8. FIRST AND REARMOST

When I was in junior high school I used to amuse myself by swinging on the rings in gym. (I was lighter then, and more foolhardy.) On one occasion I grew weary of the exercise, so at the end of one swing I let go.

It was my feeling at the time, as I distinctly remember, that I would continue my semicircular path and go swooping upward until gravity took hold; and that I would then come down light as gossamer, landing on my toes after a perfect *entrechat*.

That is not the way it happened. My path followed nearly a straight line, tangent to the semicircle of swing at the point at which I let go. I landed good and hard on one side.

After my head cleared, I stood up* and to this day that is the hardest fall I have ever taken.

I might have drawn a great deal of intellectual good out of this incident. I might have pondered on the effects of inertia; puzzled out methods of summing vectors; or deduced some facts about differential calculus.

However, I will be frank with you. What really impressed itself upon me was the fact that the force of gravity was both mighty and dangerous and that if you weren't watching every minute, it would clobber you.

Presumably, I had learned that, somewhat less drasti-

* People react oddly. After I stood up, I completely ignored my badly sprained (and possibly broken, though it later turned out not to be) right wrist and lifted my untouched left wrist to my ear. What worried me was whether my wristwatch were still running.

cally, early in life; and presumably, every human being who ever got onto his hind legs at the age of a year or less and promptly toppled, learned the same fact.

In fact, I have been told that infants have an instinctive fear of falling, and that this arose out of the survival value of having such an instinctive fear during the tree-living aeons of our simian ancestry.

We can say, then, that gravitational force is the first force with which each individual human being comes in contact. Nor can we ever manage to forget its existence, since it must be battled at every step, breath, and heartbeat. Never for one moment must we cease exerting a counterforce.

It is also comforting that this mighty and overwhelming force protects us at all times. It holds us to our planet and doesn't allow us to shoot off into space. It holds our air and water to the planet, too, for our perpetual use. And it holds the Earth itself firmly in its orbit about the Sun, so that we always get the light and warmth we need.

What with all this, it generally comes as a rather surprising shock to many people to learn that gravitation is *not* the strongest force in the universe. Suppose, for instance, we compare it with the electromagnetic force that allows a magnet to attract iron or a proton to attract an electron. (The electromagnetic force also exhibits repulsion, which gravitational force does not, but that is a detail that need not distress us at this moment.)

How can we go about comparing the relative strengths of the electromagnetic force and the gravitational force?

Let's begin by considering two objects alone in the universe. The gravitational force between them, as was discovered by Newton, can be expressed by the following equation (see also Chapter 7):

$$F_g = \frac{Gmm'}{d^2} \qquad \text{(Equation 1)}$$

where F_g is the gravitational force between the objects; m is the mass of one object; m' the mass of the other; d the distance between them; and G a universal "gravitational constant."

We must be careful about our units of measurement. If we measure mass in grams, distance in centimeters, and G in somewhat more complicated units, we will end up by determining the gravitational force in something called "dynes." (Before I'm through this chapter, the dynes will cancel out, so we need not, for present purposes, consider the dyne anything more than a one-syllable noise. It will be explained, however, in Chapter 13.)

Now let's get to work. The value of G is fixed (as far as we know) everywhere in the universe. Its value in the units I am using is 6.67×10^{-8}. If you prefer long zero-riddled decimals to exponential figures, you can express G as 0.0000000667.

Let's suppose, next, that we are considering two objects of identical mass. This means that $m = m'$, so that mm' becomes mm, or m^2. Furthermore, let's suppose the particles to be exactly 1 centimeter apart, center to center. In that case $d = 1$, and $d^2 = 1$ also. Therefore, Equation 1 simplifies to the following:

$$F_g = 0.0000000667 \, m^2 \qquad \text{(Equation 2)}$$

We can now proceed to the electromagnetic force, which we can symbolize as F_e.

Exactly one hundred years after Newton worked out the equation for gravitational forces, the French physicist Charles Augustin de Coulomb (1736-1806) was able to show that a very similar equation could be used to determine the electromagnetic force between two electrically charged objects.

Let us suppose, then, that the two objects for which we have been trying to calculate gravitational forces also carry electric charges, so that they also experience an electromagnetic force. In order to make sure that the electromagnetic force is an attracting one and is therefore directly comparable to the gravitational force, let us suppose that one object carries a positive electric charge and the other a negative one. (The principle would remain even if we used like electric charges and measured the force of electromagnetic repulsion, but why introduce distractions?)

According to Coulomb, the electromagnetic force between the two objects would be expressed by the following equation:

$$F_e = \frac{qq'}{d^2} \qquad \text{(Equation 3)}$$

where q is the charge on one object, q' on the other, and d is the distance between them.

If we let distance be measured in centimeters and electric charge in units called "electrostatic units" (usually abbreviated "esu"), it is not necessary to insert a term analogous to the gravitational constant, provided the objects are separated by a vacuum. And, of course, since I started by assuming the objects were alone in the universe, there is necessarily a vacuum between them.

Furthermore, if we use the units just mentioned, the value of the electromagnetic force will come out in dynes.

But let's simplify matters by supposing that the positive electric charge on one object is exactly equal to the negative electric charge on the other, so that $q=q'$,* which means that $qq=qq=q^2$. Again, we can allow the objects to be separated by just one centimeter, center to center, so that $d^2=1$. Consequently, Equation 3 becomes:

$$F_e = q^2 \qquad \text{(Equation 4)}$$

Let's summarize. We have two objects separated by one centimeter, center to center, each object possessing identical charge (positive in one case and negative in the other) and identical mass (no qualifications). There is both a

* We could make one of them negative to allow for the fact that one object carries a negative electric charge. Then we could say that a negative value for the electromagnetic force implies an attraction and a positive value a repulsion. However, for our purposes, none of this folderol is needed. Since electromagnetic attraction and repulsion are but opposite manifestations of the same phenomenon, we shall ignore signs.

gravitational and an electromagnetic attraction between them.

The next problem is to determine how much stronger the electromagnetic force is than the gravitational force (or how much weaker, if that is how it turns out). To do this we must determine the ratio of the forces by dividing (let us say) Equation 4 by Equation 2. The result is:

$$\frac{F_e}{F_g} = \frac{q^2}{0.0000000667\,m^2} \quad \text{(Equation 5)}$$

A decimal is an inconvenient thing to have in a denominator, but we can move it up into the numerator by taking its reciprocal (that is, by dividing it into 1). Since 1 divided by 0.0000000667 is equal to 1.5×10^7, or 15,000,000, we can rewrite Equation 5 as:

$$\frac{F_e}{F_g} = \frac{15{,}000{,}000\,q^2}{m^2} \quad \text{(Equation 6)}$$

or, still more simply, as:

$$\frac{F_e}{F_g} = 15{,}000{,}000\,(q/m)^2 \quad \text{(Equation 7)}$$

Since both F_e and F_g are measured in dynes, then in taking the ratio we find we are dividing dynes by dynes. The units, therefore, cancel out, and we are left with a "pure number." We are going to find, in other words, that one force is stronger than the other by a fixed amount; an amount that will be the same whatever units we use or whatever units an intelligent entity on the fifth planet of the star Fomalhaut wants to use. We will have, therefore, a universal constant.

In order to determine the ratio of the two forces, we see from Equation 7 that we must first determine the value of q/m; that is, the charge of an object divided by its mass. Let's consider charge first.

All objects are made up of subatomic particles of a

number of varieties. These particles fall into exactly three classes, however, with respect to electric charge:

1) Class A are those particles which, like the neutron and the neutrino, have no charge at all. Their charge is 0.

2) Class B are those particles which, like the proton and the positron, carry a positive electric charge. But all particles which carry a positive electric charge invariably carry the same quantity of positive electric charge whatever their differences in other respects (at least as far as we know). Their charge can therefore be specified as +1.

3) Class C are those particles which, like the electron and the anti-proton, carry a negative electric charge. Again, this charge is always the same in quantity. Their charge is −1.

You see, then, that an object of any size can have a net electric charge of zero, provided it happens to be made up of neutral particles and/or equal numbers of positive and negative particles.

For such an object $q=0$, and no matter how large its mass, the value of q/m is also zero. For such bodies, Equation 7 tells us, F_e/F_g is zero. The gravitational force is never zero (as long as the objects have any mass at all) and it is, therefore, under these conditions, infinitely stronger than the electromagnetic force and need be the only one considered.

This is just about the case for actual bodies. The overall net charge of the Earth and the Sun is virtually zero, and in plotting the Earth's orbit it is only necessary to consider the gravitational attraction between the two bodies.

Still, the case where $F_e=0$ and, therefore, $F_e/F_g=0$ *is* clearly only one extreme of the situation and not a particularly interesting one. What about the other extreme? Instead of an object with no charge, what about an object with maximum charge?

If we are going to make charge maximum, let's first eliminate neutral particles which add mass without charge. Let's suppose, instead, that we have a piece of matter composed exclusively of charged particles. Naturally it is of no use to include charged particles of both varieties, since then one type of charge would cancel the other and total charge would be less than maximum.

We will want one object then, composed exclusively of positively charged particles and another exclusively of negatively charged particles. We can't possibly do better than that as a general thing.

And yet while all the charged particles have identical charges of either $+1$ or -1, as the case may be, they possess different masses. What we want are charged particles of the smallest possible mass. In that case the largest possible individual charge is hung upon the smallest possible mass, and the ratio q/m is at a maximum.

It so happens that the negatively charged particle of smallest mass is the electron and the positively charged particle of smallest mass is the positron. For those bodies, the ratio q/m is greater than for any other known object (nor have we any reason, as yet, for suspecting that any object of higher q/m remains to be discovered).

Suppose, then, we start with two bodies, one of which contains a certain number of electrons and the other the same number of positrons. There will be a certain electromagnetic force between them and also a certain gravitational force.

If you triple the number of electrons in the first body and triple the number of positrons in the other, the total charge triples for each body and the total electromagnetic force, therefore, becomes 3 times 3, or 9 times greater. However, the total mass also triples for each body and the total gravitational force also becomes 3 times 3, or 9 times greater. While each force increases, they do so to an equal extent, and the ratio of the two remains the same.

In fact the ratio of the two forces remains the same, even if the charge and/or mass on one body is not equal to the charge and/or mass on the other; or if the charge and/or mass of one body is changed by an amount different from the charge in the other.

Since we are concerned only with the ratio of the two forces, the electromagnetic and the gravitational, and since this remains the same, however much the number of electrons in one body and the number of positrons in the other are changed, why bother with any but the simplest possible number—one?

In other words, let's consider a single electron and a simple positron separated by exactly 1 centimeter. This system will give us the maximum value for the ratio of electromagnetic force to gravitational force.

It so happens that the electron and the positron have equal masses. That mass, in grams (which are the mass-units we are using in this calculation) is 9.1×10^{-28} or, if you prefer, 0.00000000000000000000000000091.

The electric charge of the electron is equal to that of the positron (though different in sign). In electrostatic units (the charge-units being used in this calculation), the value is 4.8×10^{-10}, or 0.00000000048.

To get the value q/m for the electron (or the positron) we must divide the charge by the mass. If we divide 4.8×10^{-10} by 9.1×10^{-28}, we get the answer 5.3×10^{17} or 530,000,000,000,000,000.

But, as Equation 7 tells us, we must square the ratio q/m. We multiply 5.3×10^{17} by itself and obtain for $(q/m)^2$ the value of 2.8×10^{35}, or 280,000,000,000,-000,000,000,000,000,000,000,000.

Again, consulting Equation 7, we find we must multiply this number by 15,000,000, and then we finally have the ratio we are looking for. Carrying through this multiplication gives us 4.2×10^{42}, or 4,200,000,000,000,000,000,-000,000,000,000,000,000,000,000.

We can come to the conclusion, then, that the electromagnetic force is, under the most favorable conditions, over four million trillion trillion trillion times as strong as the gravitational force.

To be sure, under normal conditions there are no electron/positron systems in our surroundings, for positrons virtually do not exist. Instead our universe (as far as we know) is held together electromagnetically by electron/proton attractions. The proton is 1836 times as massive as the electron, so that the gravitational attraction is increased without a concomitant increase in electromagnetic attraction. In this case the ratio F_e/F_g is only 2.3×10^{39}.

There are two other major forces in the physical world. There is the nuclear strong interaction force which is over a hundred times as strong as even the electromagnetic force; and the nuclear weak interaction force, which is considerably weaker than the electromagnetic force. All three, however, are far, far stronger than the gravitational force.

In fact, the force of gravity—though it is the first force with which we are acquainted, and though it is always with us, and though it is the one with a strength we most thoroughly appreciate—is *by far the weakest known force in nature*. It is first and rearmost!

What makes the gravitational force *seem* so strong?

First, the two nuclear forces are short-range forces which make themselves felt only over distances about the width of an atomic nucleus. The electromagnetic force and the gravitational force are the only two long-range forces. Of these, the electromagnetic force cancels itself out (with slight and temporary local exceptions) because both an attraction and a repulsion exist.

This leaves gravitational force alone in the field.

What's more, the most conspicuous bodies in the universe happen to be conglomerations of vast mass, and we live on the surface of one of these conglomerations.

Even so, there are hints that give away the real weakness of gravitational force. Your weak muscle can lift a fifty-pound weight with the whole mass of the earth pulling, gravitationally, in the other direction. A toy magnet will lift a pin against the entire counterpull of the earth.

Oh, gravity is weak all right. But let's see if we can dramatize that weakness further.

Suppose that the Earth were an assemblage of nothing but its mass in positrons, while the Sun were an assemblage of nothing but its mass of electrons. The force of attraction between them would be vastly greater than the feeble gravitational force that holds them together now. In fact, in order to reduce the electromagnetic attraction to no more than the present gravitational one, the Earth and Sun would have to be separated by some 33,000,000,000,000,000 light-years, or about five million times the diameter of the known universe.

Or suppose you imagined in the place of the Sun a million tons of electrons (equal to the mass of a very small asteroid). And in the place of the Earth, imagine 3⅓ tons of positrons.

The electromagnetic attraction between these two insignificant masses, separated by the distance from the Earth to the Sun, would be equal to the gravitational attraction between the colossal masses of those two bodies right now.

In fact, if one could scatter a million tons of electrons on the Sun, and 3⅓ tons of positrons on the Earth, you would double the Sun's attraction for the Earth and alter the nature of Earth's orbit considerably. And if you made it electrons, both on Sun and Earth, so as to introduce a repulsion, you would cancel the gravitational attraction altogether and send old Earth on its way out of the Solar System.

Of course, all this is just paper calculation. The mere fact that electromagnetic forces are as strong as they are means that you cannot collect a significant number of like-charged particles in one place. They would repel each other too strongly.

Suppose you divided the Sun into marble-sized fragments and strewed them through the Solar System at mutual rest. Could you, by some manmade device, keep those fragments from falling together under the pull of gravity? Well, this is no greater a task than that of getting hold of a million tons of electrons and squeezing them together into a ball.

The same would hold true if you tried to separate a sizable quantity of positive charge from a sizable quantity of negative charge.

If the universe were composed of electrons and positrons as the chief charged particles, the electromagnetic force would make it necessary for them to come together. Since they are anti-particles, one being the precise reverse of the other, they would melt together, cancel each other, and go up in one cosmic flare of gamma rays.

Fortunately, the universe is composed of electrons and protons as the chief charged particles. Though their charges are exact opposites (-1 for the former and $+1$ for the latter), this is not so of other properties—such as mass, for instance. Electrons and protons are not anti-particles, in other words, and cannot cancel each other.

Their opposite charges, however, set up a strong mutual attraction that cannot, within limits, be gainsaid. An electron and a proton therefore approach closely and then maintain themselves at a wary distance, forming the hydrogen atom.

Individual protons can cling together despite electromagnetic repulsion because of the existence of a very short-range nuclear strong interaction force that sets up an attraction between neighboring protons that far overbalances the electromagnetic repulsion. This makes atoms other than hydrogen possible.

In short: nuclear forces dominate the atomic nucleus; electromagnetic forces dominate the atom itself; and gravitational forces dominate the large astronomic bodies.

The weakness of the gravitational force is a source of frustration to physicists.

The different forces, you see, make themselves felt by transfers of particles. The nuclear strong interaction force, the strongest of all, makes itself evident by transfers of pions (pi-mesons), while the electromagnetic force (next strongest) does it by the transfer of photons. An analogous particle involved in weak interactions (third strongest) has recently been reported. It is called the "w particle" and as yet the report is a tentative one.

So far, so good. It seems, then, that if gravitation is a force in the same sense that the others are, it should make itself evident by transfers of particles.

Physicists have given this particle a name, the "graviton." They have even decided on its properties, or lack of properties. It is electrically neutral and without mass. (Because it is without mass, it must travel at an unvarying velocity, that of light.) It is stable, too; that is, left to itself, it will not break down to form other particles.

So far, it is rather like the neutrino,* which is also stable, electrically, neutral, and massless (hence traveling at the velocity of light).

The graviton and the neutrino differ in some respects, however. The neutrino comes in two varieties, an electron neutrino and a muon (mu-meson) neutrino, each with its anti-particle; so there are, all told, four distinct kinds of neutrinos. The graviton comes in but one variety and is its own anti-particle. There is but one kind of graviton.

Then, too, the graviton has a spin of a type that is assigned the number 2, while the neutrino along with most other subatomic particles have spins of ½. (There are also some mesons with a spin of 0 and the photon with a spin of 1.)

The graviton has not yet been detected. It is even more elusive than the neutrino. The neutrino, while massless and chargeless, nevertheless has a measurable energy content. Its existence was first suspected, indeed, because it carried off enough energy to make a sizable gap in the bookkeeping.

But gravitons?

Well, remember that factor of 10^{42}!

An individual graviton must be trillions of trillions of trillions of times less energetic than a neutrino. Considering how difficult it was to detect the neutrino, the detection of the graviton is a problem that will *really* test the nuclear physicist.

* See Chapter 13 of my book *View from a Height,* Doubleday, 1963.

9. THE BLACK OF NIGHT

I suppose many of you are familiar with the comic strip "Peanuts." My daughter Robyn (now in the fourth grade) is very fond of it, as I am myself.

She came to me one day, delighted with a particular sequence in which one of the little characters in "Peanuts" asks his bad-tempered older sister, "Why is the sky blue?" and she snaps back, "Because it isn't green!"

When Robyn was all through laughing, I thought I would seize the occasion to maneuver the conversation in the direction of a deep and subtle scientific discussion (entirely for Robyn's own good, you understand). So I said, "Well, tell me, Robyn, why is the night sky black?"

And she answered at once (I suppose I ought to have foreseen it), "Because it isn't purple!"

Fortunately, nothing like this can ever seriously frustrate me. If Robyn won't cooperate, I can always turn, with a snarl, on the Helpless Reader. I will discuss the blackness of the night sky with *you!*

The story of the black of night begins with a German physician and astronomer, Heinrich Wilhelm Matthias Olbers, born in 1758. He practiced astronomy as a hobby, and in midlife suffered a peculiar disappointment. It came about in this fashion . . .

Toward the end of the eighteenth century, astronomers began to suspect, quit strongly, that some sort of planet must exist between the orbits of Mars and Jupiter. A team of German astronomers, of whom Olbers was one of the most important, set themselves up with the intention of dividing the ecliptic among themselves and each searching his own portion, meticulously, for the planet.

Olbers and his friends were so systematic and thorough that by rights they should have discovered the planet and received the credit of it. But life is funny (to coin a phrase). While they were still arranging the details, Giuseppe Piazzi, an Italian astronomer who wasn't looking for planets at all, discovered, on the night of January 1, 1801, a point of light which had shifted its position against the background of stars. He followed it for a period of time and found it was continuing to move steadily. It moved less rapidly than Mars and more rapidly than Jupiter, so it was very likely a planet in an intermediate orbit. He reported it as such so that it was the casual Piazzi and not the thorough Olbers who got the nod in the history books.

Olbers didn't lose out altogether, however. It seems that after a period of time, Piazzi fell sick and was unable to continue his observations. By the time he got back to the telescope the planet was too close to the Sun to be observable.

Piazzi didn't have enough observations to calculate an orbit and this was bad. It would take months for the slow-moving planet to get to the other side of the Sun and into observable position, and without a calculated orbit it might easily take years to rediscover it.

Fortunately, a young German mathematician, Karl Friedrich Gauss, was just blazing his way upward into the mathematical firmament. He had worked out something called the "method of least squares," which made it possible to calculate a reasonably good orbit from no more than three good observations of a planetary position.

Gauss calculated the orbit of Piazzi's new planet, and when it was in observable range once more there was Olbers and his telescope watching the place where Gauss's calculations said it would be. Gauss was right and, on January 1, 1802, Olbers found it.

To be sure, the new planet (named "Ceres") was a peculiar one, for it turned out to be less than 500 miles in diameter. It was far smaller than any other known planet and smaller than at least six of the satellites known at that time.

Could Ceres be all that existed between Mars and Jupiter? The German astronomers continued looking (it would be a shame to waste all that preparation) and sure enough, three more planets between Mars and Jupiter were soon discovered. Two of them, Pallas and Vesta, were discovered by Olbers. (In later years many more were discovered.)

But, of course, the big payoff isn't for second place. All Olbers got out of it was the name of a planetoid. The thousandth planetoid between Mars and Jupiter was named "Piazzia," the thousand and first "Gaussia," and the thousand and second (hold your breath, now) "Olberia."

Nor was Olbers much luckier in his other observations. He specialized in comets and discovered five of them, but practically anyone can do that. There is a comet called "Olbers' Comet" in consequence, but that is a minor distinction.

Shall we now dismiss Olbers? By no means.

It is hard to tell just what will win you a place in the annals of science. Sometimes it is a piece of interesting reverie that does it. In 1826 Olbers indulged himself in an idle speculation concerning the black of night and dredged out of it an apparently ridiculous conclusion.

Yet that speculation became "Olbers' paradox," which has come to have profound significance a century afterward. In fact, we can begin with Olbers' paradox and end with the conclusion that the only reason life exists anywhere in the universe is that the distant galaxies are receding from us.

What possible effect can the distant galaxies have on us? Be patient now and we'll work it out.

In ancient times, if any astronomer had been asked why the night sky was black, he would have answered—quite reasonably—that it was because the light of the Sun was absent. If one had then gone on to question him why the stars did not take the place of the Sun, he would have answered—again reasonably—that the stars were limited in number and individually dim. In fact, all the stars we can see would, if lumped together, be only a half-billionth as bright as the Sun. Their influence on the blackness of the night sky is therefore insignificant.

By the nineteenth century, however, this last argument had lost its force. The number of stars was tremendous. Large telescopes revealed them by the countless millions.

Of course, one might argue that those countless millions of stars were of no importance for they were not visible to the naked eye and therefore did not contribute to the light in the night sky. This, too, is a useless argument. The stars of the Milky Way are, individually, too faint to be made out, but *en masse* they make a dimly luminous belt about the sky. The Andromeda galaxy, is much farther away than the stars of the Milky Way and the individual stars that make it up are not individually visible except (just barely) in a very large telescope. Yes, *en masse,* the Andromeda galaxy is faintly visible to the naked eye. (It is, in fact, the farthest object visible to the unaided eye; so if anyone ever asks you how far you can see; tell him 2,000,000 light-years.)

In short, distant stars—no matter how distant and no matter how dim, individually—must contribute to the light of the night sky, and this contribution can even become detectable without the aid of instruments if these dim distant stars exist in sufficient density.

Olbers, who didn't know about the Andromeda galaxy, but did know about the Milky Way, therefore set about asking himself how much light ought to be expected from the distant stars altogether. He began by making several assumptions:

1. That the universe is infinite in extent.

2. That the stars are infinite in number and evenly spread throughout the universe.

3. That the stars are of uniform average brightness through all of space.

Now let's imagine space divided up into shells (like those of an onion) centering about us, comparatively thin shells compared with the vastness of space, but large enough to contain stars within them.

Remember that the amount of light that reaches us from individual stars of equal luminosity varies inversely as the square of the distance from us. In other words, if Star A and Star B are equally bright but Star A is three times as far as Star B, Star A delivers only 1/9 the light. If Star A were five times as far as Star B, Star A would deliver 1/25 the light, and so on.

This holds for our shells. The average star in a shell 2000 light-years from us would be only 1/4 as bright in appearance as the average star in a shell only 1000 light-years from us. (Assumption 3 tells us, of course, that the intrinsic brightness of the average star in both shells is the same, so that distance is the only factor we need consider.) Again, the average star in a shell 3000 light-years from us would be only 1/9 as bright in appearance as the average star in the 1000-light-year shell, and so on.

But as you work your way outward, each succeeding shell is more voluminous than the one before. Since each shell is thin enough to be considered, without appreciable error, to be the surface of the sphere made up of all the shells within, we can see that the volume of the shells increases as the surface of the spheres would—that is, as the square of the radius. The 2000-light-year shell would have four times the volume of the 1000-light-year shell. The 3000-light-year shell would have nine times the volume of the 1000-light-year shell, and so on.

If we consider the stars to be evenly distributed through space (Assumption 2), then the number of stars in any given shell is proportional to the volume of the shell. If the 2000-light-year shell is four times as voluminous as the 1000-light-year shell, it contains four times as many stars. If the 3000-light-year shell is nine times as voluminous as the 1000-light-year shell, it contains nine times as many stars, and so on.

Well, then, if the 2000-light-year shell contains four times as many stars as the 1000-light-year shell, and if each star in the former is 1/4 as bright (on the average) as each star of the latter, then the total light delivered by the

2000-light-year shell is 4 times ¼ that of the 1000-light-year shell. In other words, the 2000-light-year shell delivers just as much total light as the 1000-light-year shell. The total brightness of the 3000-light-year shell is 9 times ⅑ that of the 1000-light-year shell, and the brightness of the two shells is equal again.

In summary, if we divide the universe into successive shells, each shell delivers as much light, *in toto,* as do any of the others. And if the universe is infinite in extent (Assumption 1) and therefore consists of an infinite number of shells, the stars of the universe, however dim they may be individually, ought to deliver an infinite amount of light to the Earth.

The one catch, of course, is that the nearer stars may block the light of the more distant stars.

To take this into account, let's look at the problem in another way. In no matter which direction one looks, the eye will eventually encounter a star, if it is true they are infinite in number and evenly distributed in space (Assumption 2). The star may be individually invisible, but it will contribute its bit of light and will be immediately adjoined in all directions by other bits of light.

The night sky would then not be black at all but would be an absolutely solid smear of starlight. So would the day sky be an absolutely solid smear of starlight, with the Sun itself invisible against the luminous background.

Such a sky would be roughly as bright as 150,000 suns like ours, and do you question that under those conditions life on Earth would be impossible?

However, the sky is *not* as bright as 150,000 suns. The night sky *is* black. Somewhere in the Olbers' paradox there is some mitigating circumstance or some logical error.

Olbers himself thought he found it. He suggested that space was not truly transparent; that it contained clouds of dust and gas which absorbed most of the starlight, allowing only an insignificant fraction to reach the Earth.

That sounds good, but it is no good at all. There are indeed dust clouds in space but if they absorbed all the starlight that fell upon them (by the reasoning of Olbers' paradox) then their temperature would go up until they grew hot enough to be luminous. They would, eventually, emit as much light as they absorb and the Earth sky would still be star-bright over all its extent.

But if the logic of an argument is faultless and the conclusion is still wrong, we must investigate the assumptions. What about Assumption 2, for instance? Are the stars indeed infinite in number and evenly spread throughout the universe?

Even in Olbers' time there seemed reason to believe this assumption to be false. The German-English astronomer William Herschel made counts of stars of different brightness. He assumed that, *on the average,* the dimmer stars were more distant than the bright ones (which follows from Assumption 3) and found that the density of the stars in space fell off with distance.

From the rate of decrease in density in different directions, Herschel decided that the stars made up a lens-shaped figure. The long diameter, he decided, was 150 times the distance from the Sun to Arcturus (or 6000 light-years, we would now say), and the whole conglomeration would consist of 100,000,000 stars.

This seemed to dispose of Olbers' paradox. If the lens-shaped conglomerate (now called the Galaxy) truly contained all the stars in existence, then Assumption 2 breaks down. Even if we imagined space to be infinite in extent outside the Galaxy (Assumption 1), it would contain no stars and would contribute no illumination. Consequently, there would be only a finite number of star-containing shells and only a finite (and not very large) amount of illumination would be received on Earth. That would be why the night sky is black.

The estimated size of the Galaxy has been increased since Herschel's day. It is now believed to be 100,000 light-years in diameter, not 6000; and to contain 150,000,000,000 stars, not 100,000,000. This change, however, is not crucial; it still leaves the night sky black.

In the twentieth century Olbers' paradox came back to life, for it came to be appreciated that there were indeed stars outside the Galaxy.

The foggy patch in Andromeda had been felt throughout the nineteenth century to be a luminous mist that formed part of our own Galaxy. However, other such patches of mist (the Orion Nebula, for instance) contained stars that lit up the mist. The Andromeda patch, on the other hand, seemed to contain no stars but to glow of itself.

Some astronomers began to suspect the truth, but it wasn't definitely established until 1924, when the American astronomer Edwin Powell Hubble turned the 100-inch telescope on the glowing mist and was able to make out separate stars in its outskirts. These stars were individually so dim that it became clear at once that the patch must be hundreds of thousands of light-years away from us and far outside the Galaxy. Furthermore, to be seen, as it was, at that distance, it must rival in size our entire Galaxy and be another galaxy in its own right.

And so it is. It is now believed to be over 2,000,000 light-years from us and to contain at least 200,000,000,000 stars. Still other galaxies were discovered at vastly greater distances. Indeed, we now suspect that within the observable universe there are at least 100,000,000,000 galaxies, and the distance of some of them has been estimated as high as 6,000,000,000 light-years.

Let us take Olbers' three assumptions then and substitute the word "galaxies" for "stars" and see how they sound.

Assumption 1, that the universe is infinite, sounds good. At least there is no sign of an end even out to distances of billions of light-years.

Assumption 2, that *galaxies* (not stars) are infinite in number and evenly spread throughout the universe, sounds good, too. At least they are evenly distributed for as far out as we can see, and we can see pretty far.

Assumption 3, that *galaxies* (not stars) are of uniform average brightness throughout space, is harder to handle. However, we have no reason to suspect that distant galaxies are consistently large or smaller than nearby ones, and if the galaxies come to some uniform average size and star-content, then it certainly seems reasonable to suppose they are uniformly bright as well.

Well, then, why is the night sky black? We're back to that.

Let's try another tack. Astronomers can determine whether a distant luminous object is approaching us or receding from us by studying its spectrum (that is, its light as spread out in a rainbow of wavelengths from short-wavelength violet to long-wavelength red).

The spectrum is crossed by dark lines which are in a fixed position if the object is motionless with respect to us. If the object is approaching us, the lines shift toward the violet. If the object is receding from us, the lines shift toward the red. From the size of the shift astronomers can determine the velocity of approach or recession.

In the 1910s and 1920s the spectra of some galaxies (or bodies later understood to be galaxies) were studied, and except for one or two of the very nearest, all are receding from us. In fact, it soon became apparent that the farther galaxies are receding more rapidly than the nearer ones. Hubble was able to formulate what is now called "Hubble's Law" in 1929. This states that the velocity of recession of a galaxy is proportional to its distance from us. If Galaxy A is twice as far as Galaxy B, it is receding at twice the velocity. The farthest observed galaxy, 6,000,000,000 light-years from us, is receding at a velocity half that of light.

The reason for Hubble's Law is taken to lie in the expansion of the universe itself—an expansion which can be made to follow from the equations set up by Einstein's General Theory of Relativity (which, I hereby state firmly, I will *not* go into).

Given the expansion of the universe, now, how are Olbers' assumptions affected?

If, at a distance of 6,000,000,000 light-years a galaxy recedes at half the speed of light, then at a distance of 12,000,000,000 light-years a galaxy ought to be receding at the speed of light (if Hubble's Law holds). Surely, further distances are meaningless, for we cannot have velocities greater than that of light. Even if that were possible, no light, or any other "message" could reach us from such a more-distant galaxy and it would not, in effect, be in our universe. Consequently, we can imagine the universe to be finite after all, with a "Hubble radius" of some 12,000,000,000 light-years.

But that doesn't wipe out Olbers' paradox. Under the requirements of Einstein's theories, as galaxies move faster and faster relative to an observer, they become shorter and shorter in the line of travel and take up less and less space, so that there is room for larger and larger numbers of galaxies. In fact, even in a finite universe, with a radius of 12,000,000,000 light-years, there might still be an infinite number of galaxies; almost all of them (paper-thin) existing in the outermost few miles of the Universe-sphere.

So Assumption 2 stands even if Assumption 1 does not; and Assumption 2, by itself, can be enough to insure a star-bright sky.

But what about the red shift?

Astronomers measure the red shift by the change in position of the spectral lines, but those lines move only because the entire spectrum moves. A shift to the red is a shift in the direction of lesser energy. A receding galaxy delivers less radiant energy to the Earth than the same galaxy would deliver if it were standing still relative to us—just because of the red shift. The faster a galaxy recedes the less radiant energy it delivers. A galaxy receding at the speed of light delivers no radiant energy at all no matter how bright it might be.

Thus, Assumption 3 fails! It would hold true if the universe were static, but not if it is expanding. Each succeeding shell in an expanding universe delivers less light than the one within because its content of galaxies is successively farther from us; is subjected to a successively greater red shift; and falls short, more and more, of the expected radiant energy it might deliver.

And because Assumption 3 fails, we receive only a

finite amount of energy from the universe and the night sky is black.

According to the most popular models of the universe, this expansion will always continue. It may continue without the production of new galaxies so that, eventually, billions of years hence, our Galaxy (plus a few of its neighbors, which together make up the "local cluster" of galaxies) will seem alone in the universe. All the other galaxies will have receded too far to detect. Or new galaxies may continuously form so that the universe will always seem full of galaxies, despite its expansion. Either way, however, expansion will continue and the night sky will remain black.

There is another suggestion, however, that the universe oscillates; that the expansion will gradually slow down until the universe comes to a moment of static pause, then begins to contract again, faster and faster, till it tightens at last into a small sphere that explodes and brings about a new expansion.

If so, then as the expansion slows the dimming effect of the red shift will diminish and the night sky will slowly brighten. By the time the universe is static the sky will be uniformly star-bright as Olbers' paradox required. Then, once the universe starts contracting, there will be a "violet-shift" and the energy delivered will increase so that the sky will become far brighter and still brighter.

This will be true not only for the Earth (if it still existed in the far future of a contracting universe) but for any body of any sort in the universe. In a static or, worse still, a contracting universe there could, by Olbers' paradox, be no cold bodies, no solid bodies. There would be uniform high temperatures everywhere—in the millions of degrees, I suspect—and life simply could not exist.

So I get back to my earlier statement. The reason there is life on Earth, or anywhere in the universe, is simply that the distant galaxies are moving away from us.

In fact, now that we know the ins and outs of Olbers' paradox, might we, do you suppose, be able to work out the recession of the distant galaxies as a necessary consequence of the blackness of the night sky? Maybe we could amend the famous statement of the French philosopher René Descartes.

He said, "I think, therefore I am!"

And we could add: "I am, therefore the universe expands!"

10. A GALAXY AT A TIME

Four or five years ago there was a small fire at a school two blocks from my house. It wasn't much of a fire, really, producing smoke and damaging some rooms in the basement, but nothing more. What's more, it was outside school hours so that no lives were in danger.

Nevertheless, as soon as the first piece of fire apparatus was on the scene the audience had begun to gather. Every idiot in town and half the idiots from the various contiguous towns came racing down to see the fire. They came by auto and by oxcart, on bicycle and on foot. They came with girl friends on their arms, with aged parents on their shoulders, and with infants at the breast.

They parked all the streets solid for miles around and after the first fire engine had come on the scene nothing more could have been added to it except by helicopter.

Apparently this happens every time. At every disaster, big or small, the two-legged ghouls gather and line up shoulder to shoulder and chest to back. They do this, it seems, for two purposes: a) to stare goggle-eyed and slack-jawed at destruction and misery, and b) to prevent the approach of the proper authorities who are attempting to safeguard life and property.

Naturally, I wasn't one of those who rushed to see the fire and I felt very self-righteously noble about it. However (since we are all friends), I will confess that this is not necessarily because I am free of the destructive instinct. It's just that a messy little fire in a basement isn't *my* idea of destruction; or a good, roaring blaze at the munitions dump, either.

If a star were to blow up, *then* we might have something.

Come to think of it, my instinct for destruction must be well developed after all, or I wouldn't find myself so fascinated by the subject of supernovas, those colossal stellar explosions.

Yet in thinking of them, I have, it turns out, been a piker. Here I've been assuming for years that a supernova was the grandest spectacle the universe had to offer (provided you were standing several dozen light-years away) but, thanks to certain 1963 findings, it turns out that a supernova taken by itself is not much more than a two-inch firecracker.

This realization arose out of radio astronomy. Since World War II, astronomers have been picking up microwave (very short radio-wave) radiation from various parts of the sky, and have found that some of it comes from our own neighborhood. The Sun itself is a radio source and so are Jupiter and Venus.

The radio sources of the Solar System, however, are virtually insignificant. We would never spot them if we weren't right here with them. To pick up radio waves across the vastness of stellar distances we need something better. For instance, one radio source from beyond the Solar System is the Crab Nebula. Even after its radio waves have been diluted by spreading out for five thousand light-years before reaching us, we can still pick up what remains and impinges upon our instruments. But then the Crab Nebula represents the remains of a supernova that blew itself to kingdom come—the first light of the explosion reaching the Earth about 900 years ago.

But a great number of radio sources lie outside our Galaxy altogether and are millions and even billions of light-years distant. *Still* their radio-wave emanations can be detected and so they must represent energy sources that shrink mere supernovas to virtually nothing.

For instance, one particularly strong source turned out, on investigation, to arise from a galaxy 200,000,000 light-years away. Once the large telescopes zeroed in on that galaxy it turned out to be distorted in shape. After closer study it became quite clear that it was not a galaxy at all, but *two* galaxies in the process of collision.

When two galaxies collide like that, there is little likelihood of actual collisions between stars (which are too small and too widely spaced). However, if the galaxies possess clouds of dust (and many galaxies, including our own, do), these clouds will collide and the turbulence of the collision will set up radio-wave emission, as does the turbulence (in order of decreasing intensity) of the gases of the Crab Nebula, of our Sun, of the atmosphere of Jupiter, and of the atmosphere of Venus.

But as more and more radio sources were detected and pinpointed, the number found among the far-distant galaxies seemed impossibly high. There might be occasional collisions among galaxies but it seemed most unlikely that there could be enough collisions to account for all those radio sources.

Was there any other possible explanation? What was needed was some cataclysm just as vast and intense as that represented by a pair of colliding galaxies, but one that involved a single galaxy. Once freed from the necessity of supposing collisions we can explain any number of radio sources.

But what can a single galaxy do alone, without the help of a sister galaxy?

Well, it can explode.

But how? A galaxy isn't really a single object. It is simply a loose aggregate of up to a couple of hundred billion stars. These stars can explode individually, but how can we have an explosion of a whole galaxy at a time?

To answer that, let's begin by understanding that a galaxy isn't really as loose an aggregation as we might tend to think. A galaxy like our own may stretch out 100,000 light-years in its longest diameter, but most of that consists of nothing more than a thin powdering of stars—thin enough to be ignored. We happen to live in this thinly starred outskirt of our own Galaxy so we accept that as the norm, but it isn't.

The nub of a galaxy is its nucleus, a dense packet of

stars roughly spherical in shape and with a diameter of, say, 10,000 light-years. Its volume is then 525,000,000,000 cubic light-years, and if it contains 100,000,000,000 stars, that means there is 1 star per 5.25 cubic light-years.

With stars massed together like that, the average distance between stars in the galactic nucleus is 1.7 light-years—but that's the average over the entire volume. The density of star numbers in such a nucleus increases as one moves toward the center, and I think it is entirely fair to expect that toward the center of the nucleus, stars are not separated by more than half a light-year.

Even half a light-year is something like 3,000,000,000,000 miles or 400 times the extreme width of Pluto's orbit, so that the stars aren't actually *crowded;* they're not likely to be colliding with each other, and yet . . .

Now suppose that, somewhere in a galaxy, a supernova lets go.

What happens?

In most cases, nothing (except that one star is smashed to flinders). If the supernova were in a galactic suburb—in our own neighborhood, for instance—the stars would be so thinly spread out that none of them would be near enough to pick up much in the way of radiation. The incredible quantities of energy poured out into space by such a supernova would simply spread and thin out and come to nothing.

In the center of a galactic nucleus, the supernova is not quite as easy to dismiss. A good supernova at its height is releasing energy at nearly 10,000,000,000 times the rate of our Sun. An object five light-years away would pick up a tenth as much energy per second as the Earth picks up from the Sun. At half a light-year from the supernova it would pick up ten times as much energy per second as Earth picks up from the Sun.

This isn't good. If a supernova let go five light-years from us we would have a year of bad heat problems. If it were half a light-year away I suspect there would be little left of earthly life. However, don't worry. There is only one star-system within five light-years of us and it is not the kind that can go supernova.

But what about the effects on the stars themselves? If our Sun were in the neighborhood of a supernova it would be subjected to a bath of energy and its own temperature would have to go up. After the supernova is done, the Sun would seek its own equilibrium again and be as good as before (though life on its planets may not be). However, in the process, it would have increased its fuel consumption in proportion to the fourth power of its absolute temperature. Even a small rise in temperature might lead to a surprisingly large consumption of fuel.

It is by fuel consumption that one measures a star's age. When the fuel supply shrinks low enough, the star expands into a red giant or explodes into a supernova. A distant supernova by warming the Sun slightly for a year might cause it to move a century, or ten centuries closer to such a crisis. Fortunately, our Sun has a long lifetime ahead of it (several billion years), and a few centuries or even a million years would mean little.

Some stars, however, cannot afford to age even slightly. They are already close to that state of fuel consumption which will lead to drastic changes, perhaps even supernovahood. Let's call such stars, which are on the brink, presupernovas. How many of them would there be per galaxy?

It has been estimated that there are an average of 3 supernovas per century in the average galaxy. That means that in 33,000,000 years there are about a million supernovas in the average galaxy. Considering that a galactic life span may easily be a hundred billion years, any star that's only a few million years removed from supernovahood may reasonably well be said to be on the brink.

If, out of the hundred billion stars in an average galactic nucleus, a million stars are on the brink, then 1 star out of 100,000 is a pre-supernova. This means that presupernovas within galactic nuclei are separated by average distances of 80 light-years. Toward the center of the nucleus, the average distance of separation might be as low as 25 light-years.

But even at 25 light-years, the light from a supernova would be only 1/250 that which the Earth receives from the Sun, and its effect would be trifling. And, as a matter of fact, we frequently see supernovas light up one galaxy or another and nothing happens. At least, the supernova slowly dies out and the galaxy is then as it was before.

However, if the average galaxy has 1 pre-supernova in every 100,000 stars, particular galaxies may be poorer than that in supernovas—or richer. An occasional galaxy may be particularly rich and 1 star out of every 1000 may be a pre-supernova.

In such a galaxy, the nucleus would contain 100,000,000 pre-supernovas, separated by an average distance of 17 light-years. Toward the center, the average separation might be no more than 5 light-years. If a supernova lights up a pre-supernova only 5 light-years away it will shorten its life significantly, and if that supernova had been a thousand years from explosion before, it might be only two months from explosion afterward.

Then, when it lets go, a more distant pre-supernova which has had its lifetime shortened, but not so drastically, by the first, may have its lifetime shortened again by the second and closer supernova, and after a few months *it* blasts.

On and on like a bunch of tumbling dominoes this would go, until we end up with a galaxy in which not a single supernova lets bang, but several million perhaps, one after the other.

There is the galactic explosion. Surely such a tumbling of dominoes would be sufficient to give birth to a coruscation of radio waves that would still be easily detectable even after it had spread out for a billion light-years.

Is this just speculation? To begin with, it was, but in late 1963 some observational data made it appear to be more than that.

It involves a galaxy in Ursa Major which is called M82 because it is number 82 on a list of objects in the heavens prepared by the French astronomer Charles Messier about two hundred years ago.

Messier was a comet-hunter and was always looking through his telescope and thinking he had found a comet and turning handsprings and then finding out that he had been fooled by some foggy object which was always there and was *not* a comet.

Finally, he decided to map each of 101 annoying objects that were foggy but were not comets so that others would not be fooled as he was. It was that list of annoyances that made his name immortal.

The first on his list, M1, is the Crab Nebula. Over two dozen are globular clusters (spherical conglomerations of densely strewn stars), M13 being the Great Hercules Cluster, which is the largest known. Over thirty members of his list are galaxies, including the Andromeda Galaxy (M31) and the Whirlpool Galaxy (M51). Other famous objects on the list are the Orion Nebula (M42), the Ring Nebula (M57), and the Owl Nebula (M97).

Anyway, M82 is a galaxy about 10,000,000 light-years from Earth which aroused interest when it proved to be a strong radio source. Astronomers turned the 200-inch telescope upon it and took pictures through filters that blocked all light except that coming from hydrogen ions. There was reason to suppose that any disturbances that might exist would show up most clearly among the hydrogen ions.

They did! A three-hour exposure revealed jets of hydrogen up to a thousand light-years long, bursting out of the galactic nucleus. The total mass of hydrogen being shot out was the equivalent of at least 5,000,000 average stars. From the rate at which the jets were traveling and the distance they had covered, the explosion must have taken place about 1,500,000 years before. (Of course, it takes light ten million years to reach us from M82, so that the explosion took place 11,500,000 years ago, Earth-time—just at the beginning of the Pleistocene Epoch.)

M82, then, is the case of an exploding galaxy. The energy expended is equivalent to that of five million supernovas formed in rapid succession, like uranium atoms undergoing fission in an atomic bomb—though on a vastly greater scale, to be sure. I feel quite certain that if there had been any life anywhere in that galactic nucleus, there isn't any now.

In fact, I suspect that even the outskirts of the galaxy may no longer be examples of prime real estate.

Which brings up a horrible thought ... Yes, you guessed it!

What if the nucleus of our own dear Galaxy explodes? It very likely won't, of course (I don't want to cause fear and despondency among the Gentle Readers), for exploding galaxies are probably as uncommon among galaxies as exploding stars are among stars. Still, if it's not going to happen, it is all the more comfortable then, as an intellectual exercise, to wonder about the consequences of such an explosion.

To begin with, we are not in the nucleus of our Galaxy but far in the outskirts and in distance there is a modicum of safety. This is especially so since between ourselves and the nucleus are vast clouds of dust that will effectively screen off any visible fireworks.

Of course, the radio waves would come spewing out, through dust and all, and this would probably ruin radio astronomy for millions of years by blanking out everything else. Worse still would be the cosmic radiation that might rise high enough to become fatal to life. In other words, we might be caught in the fallout of that galactic explosion.

Suppose, though, we put cosmic radiation to one side, since the extent of its formation is uncertain and since consideration of its presence would be depressing to the spirits. Let's also abolish the dust clouds with a wave of the speculative hand.

Now we can see the nucleus. What does it look like without an explosion?

Considering the nucleus to be 10,000 light-years in diameter and 30,000 light-years away from us, it would be visible as a roughly spherical area about 20° in diameter. When entirely above the horizon it would make up a patch about $\frac{1}{65}$ of the visible sky.

Its total light would be about 30 times that given off by

Venus at its brightest, but spread out over so large an area it would look comparatively dim. An area of the nucleus equal in size to the full Moon would have an average brightness only 1/200,000 of the full Moon.

It would be visible then as a patch of luminosity broadening out of the Milky Way in the constellation of Sagittarius, distinctly brighter than the Milky Way itself; brightest at the center, in fact, and fading off with distance from the center.

But what if the nucleus of our Galaxy exploded? The explosion would take place, I feel certain, in the center of the nucleus, where the stars were thickest and the effect of the pre-supernova on its neighbors would be most marked. Let us suppose that 5,000,000 supernovas are formed, as in M82.

If the nucleus has pre-supernovas separated by 5 light-years in its central regions (as estimated earlier in the chapter, for galaxies capable of explosion), then 5,000,000 pre-supernovas would fit into a sphere about 850 light-years in diameter. At a distance of 30,000 light-years, such a sphere would appear to have a diameter of 1.6°, which is a little more than three times the apparent diameter of the full Moon. We would therefore have an excellent view.

Once the explosion started, supernova ought to follow supernova at an accelerating rate. It would be a chain reaction.

If we were to look back on that vast explosion millions of years later, we could say (and be roughly correct) that the center of the nucleus had all exploded at once. But this is only roughly correct. If we actually watch the explosion in process, we will find it will take considerable time, thanks entirely to the fact that light takes considerable time to travel from one star to another.

When a supernova explodes, it can't affect a neighboring presupernova (5 light-years away, remember) until the radiation of the first star reaches the second—and that would take 5 years. If the second star was on the far side of the first (with respect to ourselves), an additional 5 years would be lost while the light traveled back to the vicinity of the first. We would therefore see the second supernova 10 years later than the first.

Since a supernova will not remain visible to the naked eye for more than a year or so even under the best conditions (at the distance of the Galactic nucleus), the second supernova would not be visible until long after the first had faded off to invisibility.

In short, the 5,000,000 supernovas, forming in a sphere 850 light-years in diameter, would be seen by us to appear over a stretch of time equal to roughly a thousand years. If the explosions started at the near edge of that sphere so that radiation had to travel away from us and return to set off other supernovas, the spread might easily be 1500 years. If it started at the far end and additional explosions took place as the light of the original explosion passed the presupernovas *en route* to ourselves, the time-spread might be considerably less.

On the whole, the chances are that the Galactic nucleus would begin to show individual twinkles. At first there might be only three or four twinkles a decade, but then, as the decades and centuries passed, there would be more and more until finally there might be several hundred visible at one time. And finally, they would all go out and leave behind dimly glowing gaseous turbulence.

How bright will the individual twinkles be? A single supernova can reach a maximum absolute magnitude of −17. That means if it were at a distance of 10 parsecs (32.5 light-years) from ourselves, it would have an apparent magnitude of −17, which is 1/10,000 the brightness of the Sun.

At a distance of 30,000 light-years, the apparent magnitude of such a supernova would decline by 15 magnitudes. The apparent magnitude would now be −2, which is about the brightness of Jupiter at its brightest.

This is quite a startling statistic. At the distance of the nucleus, no ordinary star can be individually seen with the naked eye. The hundred billion stars of the nucleus just make up a luminous but featureless haze under ordinary conditions. For a single star, at that distance, to fire up to the apparent brightness of Jupiter is simply colossal. Such a supernova, in fact, burns with a tenth the light intensity of an entire non-exploding galaxy such as ours.

Yet it is unlikely that every supernova forming will be a supernova of maximum brilliance. Let's be conservative and suppose that the supernovas will be, on the average, two magnitudes below the maximum. Each will then have a magnitude of 0, about that of the star Arcturus.

Even so, the "twinkles" would be prominent indeed. If humanity were exposed to such a sight in the early stages of civilization, they would never make the mistake of thinking that the heavens were eternally fixed and unchangeable. Perhaps the absence of that particular misconception (which, in actual fact, mankind labored under until early modern times) might have accelerated the development of astronomy.

However, we can't see the Galactic nucleus and that's that. Is there anything even faintly approaching such a multi-explosion that we *can* see?

There's one conceivable possibility. Here and there, in our Galaxy, are to be found globular clusters. It is estimated there are about 200 of these per galaxy. (About a hundred of our own clusters have been observed, and the other hundred are probably obscured by the dust clouds.)

These globular clusters are like detached bits of galactic nuclei, 100 light-years or so in diameter and containing from 100,000 to 10,000,000 stars—symmetrically scattered about the galactic center.

The largest known globular cluster is the Great Hercules Cluster, M13, but it is not the closest. The nearest globular cluster is Omega Centauri, which is 22,000 light-years from us and is clearly visible to the naked eye as an object of the fifty magnitude. It is only a point of light to the naked eye, however, for at that distance even a diameter of 100 light-years covers an area of only about 1.5 minutes of arc in diameter.

Now let us say that Omega Centauri contained 10,000 pre-supernovas and that every one of these exploded at their earliest opportunity. There would be fewer twinkles altogether, but they would appear over a shorter time interval and would be, individually, twice as bright.

It would be a perfectly ideal explosion for it would be unobscured by dust clouds; it would be small enough to be quite safe; and large enough to be sufficiently spectacular for anyone.

And yet, now that I've worked up my sense of excitement over the spectacle, I must admit that the chances of viewing an explosion in Omega Centauri are just about nil. And even if it happened, Omega Centauri is not visible in New England and I would have to travel quite a bit southward if I expected to see it high in the sky in full glory—and I don't like to travel.

Hmm . . . Oh well, anyone for a neighborhood fire?

Part II

OF OTHER THINGS

11. FORGET IT

The other day I was looking through a new textbook on biology *(Biological Science: An Inquiry into Life,* written by a number of contributing authors and published by Harcourt, Brace & World, Inc. in 1963). I found it fascinating.

Unfortunately, though, I read the Foreword first (yes I'm one of *that* kind) and was instantly plunged into the deepest gloom. Let me quote from the first two paragraphs:

"With each new generation our fund of scientific knowledge increases fivefold . . . At the current rate of scientific advance, there is about four times as much significant biological knowledge today as in 1930, and about sixteen times as much as in 1900. By the year 2000, at this rate of increase, there will be a hundred times as much biology to 'cover' in the introductory course as at the beginning of the century."

Imagine how this affects me. I am a professional "keeper-upper" with science and in my more manic, ebullient, and carefree moments, I even think I succeed fairly well.

Then I read something like the above-quoted passage and the world falls about my ears. I *don't* keep up with science. Worse, I *can't* keep up with it. Still worse, I'm falling farther behind every day.

And finally, when I'm all through sorrowing for myself, I devote a few moments to worrying about the world generally. What is going to become of Homo sapiens? We're going to smarten ourselves to death. After a while, we will all die of pernicious education, with our brain cells crammed to indigestion with facts and concepts, and with blasts of information exploding out of our ears.

But then, as luck would have it, the very day after I read the Foreword to *Biological Science* I came across an old, old book entitled *Pike's Arithmetic*. At least that is the name on the spine. On the title page it spreads itself a bit better, for in those days titles were *titles*. It goes "A New and Complete System of Arithmetic Composed for the Use of the Citizens of the United States," by Nicolas Pike, A.M. It was first published in 1785, but the copy I have is only the "Second Edition, Enlarged," published in 1797.

It is a large book of over 500 pages, crammed full of small print and with no relief whatever in the way of illustrations or diagrams. It is a solid slab of arithmetic except for small sections at the very end that introduce algebra and geometry.

I was amazed. I have two children in grade school (and once I was in grade school myself), and I know what arithmetic books are like these days. They are nowhere near as large. They can't possibly have even one-fifth the wordage of Pike.

Can it be that we are leaving anything out?

So I went through Pike and, you know, we *are* leaving something out. And there's nothing wrong with that. The trouble is we're not leaving *enough* out.

On page 19, for instance, Pike devotes half a page to a listing of numbers as expressed in Roman numerals, extending the list to numbers as high as five hundred thousand.

Now Arabic numerals reached Europe in the High Middle Ages, and once they came on the scene the Roman numerals were completely outmoded. They lost all possible use, so infinitely superior was the new Arabic notation. Until then who knows how many reams of paper were required to explain methods for calculating with Roman numerals. Afterward the same calculations could be performed with a hundredth of the explanation. No knowledge was lost—only inefficient rules.

And yet five hundred years after the deserved death of the Roman numerals, Pike still included them and expected his readers to be able to translate them into Arabic numerals and vice versa even though he gave no instructions for how to manipulate them. In fact, nearly two hundred years after Pike, the Roman numerals are still being taught. My little daughter is learning them now.

But why? Where's the need? To be sure, you will find Roman numerals on cornerstones and gravestones, on clockfaces and on some public buildings and documents, but it isn't used for any need at all. It is used for show, for status, for antique flavor, for a craving for some kind of phony classicism.

I dare say there are some sentimental fellows who feel that knowledge of the Roman numerals is a kind of gateway to history and culture; that scrapping them would be like knocking over what is left of the Parthenon, but I have no patience with such mawkishness. We might as well suggest that everyone who learns to drive a car be required to spend some time at the wheel of a Model-T Ford so he could get the flavor of early cardom.

Roman numerals? Forget it!—And make room instead for new and valuable material.

But do we dare forget things? Why not? We've forgotten much; more than you imagine. Our troubles stem not from the fact that we've forgotten, but that we remember too well; we don't forget enough.

A great deal of Pike's book consists of material we have imperfectly forgotten. That is why the modern arithmetic book is shorter than Pike. And if we could but perfectly forget, the modern arithmetic book could grow still shorter.

For instance, Pike devotes many pages to tables—presumably important tables that he thought the reader ought to be familiar with. His fifth table is labeled "cloth measure."

Did you know that 2¼ inches make a "nail"? Well, they do. And 16 nails make a yard; while 12 nails make an ell.

No, wait a while. Those 12 nails (27 inches) makes a *Flemish* ell. It takes 20 nails (45 inches) to make an English ell, and 24 nails (54 inches) to make a French ell. Then, 16 nails plus 1⅕ inches (37⅕ inches) make a Scotch ell.

Now if you're going to be in the business world and import and export cloth, you're going to have to know all those ells—unless you can figure some way of getting the ell out of business.

Furthermore, almost every piece of goods is measured in its own units. You speak of a firkin of butter, a punch of prunes, a fother of lead, a stone of butcher's meat, and so on. Each of these quantities weighs a certain number of pounds (avoirdupois pounds, but there are also troy pounds and apothecary pounds and so on), and Pike carefully gives all the equivalents.

Do you want to measure distances? Well, how about this: 7 92/100 inches make 1 link; 25 links make 1 pole; 4 poles make 1 chain; 10 chains make 1 furlong; and 8 furlongs make 1 mile.

Or do you want to measure ale or beer—a very common line of work in Colonial times. You have to know the language, of course. Here it is: 2 pints make a quart and 4 quarts make a gallon. Well, we still know that much anyway.

In Colonial times, however, a mere gallon of beer or ale was but a starter. That was for infants. You had to know how to speak of man-sized quantities. Well, 8 gallons make a firkin—that is, it makes a "firkin of ale in London." It takes, however, 9 gallons to make "a firkin of beer in London." The intermediate quantity, 8½ gallons, is marked down as "a firkin of ale or beer"—presumably outside of the environs of London where the provincial citizens were less finicky in distinguishing between the two.

But we go on: 2 firkins (I suppose the intermediate kind, but I'm not sure) makes a kilderkin and 2 kilderkins make a barrel. Then 1½ barrels make 1 hogshead; 2 barrels make a puncheon; and 3 barrels make a butt.

Have you got all that straight?

But let's try dry measure in case your appetite has been sharpened for something still better.

Here, 2 pints make a quart and 2 quarts make a pottle. (No, not bottle, *pottle.* Don't tell me you've never heard of a pottle!) But let's proceed.

Next, 2 pottles make a gallon, 2 gallons make a peck, and 4 pecks make a bushel. (Long breath now.) Then 2 bushels make a strike, 2 strikes make a coom, 2 cooms make a quarter, 4 quarters make a chaldron (though in the demanding city of London, it takes 4½ quarters to make a chaldron). Finally, 5 quarters make a wey and 2 weys make a last.

I'm not making this up. I'm copying it right out of Pike, page 48.

Were people who were studying arithmetic in 1797 expected to memorize all this? Apparently, yes, because Pike spends a lot of time on compound addition. That's right, *compound* addition.

You see, the addition you consider addition is just "simple addition." Compound addition is something stronger and I will now explain it to you.

Suppose you have 15 apples, your friend has 17 apples, and a passing stranger has 19 apples and you decide to make a pile of them. Having done so, you wonder how many you have altogether. Preferring not to count, you draw upon your college education and prepare to add $15 + 17 + 19$. You begin with the units column and find that $5+7+9=21$. You therefore divide 21 by 10 and find the quotient is 2 plus a remainder of 1, so you put down the remainder, 1, and carry the quotient 2 into the tens col——

I seem to hear loud yells from the audience. "What is all this? comes the fevered demand. "Where does this 'divide by 10' jazz come from?"

Ah, Gentle Readers, but this is exactly what you do whenever you add. It is only that the kindly souls who devised our Arabic system of numeration based it on the number 10 in such a way that when any two-digit number is divided by 10, the first digit represents the quotient and the second the remainder.

For that reason, having the quotient and remainder in our hands without dividing, we can add automatically. If the units column adds up to 21, we put down 1 and carry 2; if it had added up to 57, we would have put down 7 and carried 5, and so on.

The only reason this works, mind you, is that in adding a set of figures, each column of digits (starting from the right and working leftward) represents a value ten times as great as the column before. The rightmost column is units, the one to its left is tens, the one to its left is hundreds, and so on.

It is this combination of a number system based on ten and a value ratio from column to column of ten that makes addition very simple. It is for this reason that it is, as Pike calls it, "simple addition."

Now suppose you have 1 dozen and 8 apples, your friend has 1 dozen and 10 apples, and a passing stranger has 1 dozen and 9 apples. Make a pile of those and add them as follows:

1 dozen	8 units
1 dozen	10 units
1 dozen	9 units

Since $8+10+9=27$, do we put down 7 and carry 2? Not at all! The ratio of the "dozens" column to the "units" column is not 10 but 12, since there are 12 units to a dozen. And since the number system we are using is based on 10 and not on 12, we can no longer let the digits do our thinking for us. We have to go long way round.

If $8+10+9=27$, we must divide that sum by the ratio of the value of the columns; in this case, 12. We find that 27 divided by 12 gives a quotient of 2 plus a remainder of 3, so we put down 3 and carry 2. In the dozens column we get $1+1+1+2=5$. Our total therefore is 5 dozen and 3 apples.

Whenever a ratio of other than 10 is used so that you have to make actual divisions in adding, you have "compound addition." You must indulge in compound addition if you try to add 5 pounds 12 ounces and 6 pounds 8 ounces, for there are 16 ounces to a pound. You are stuck again if you add 3 yards 2 feet 6 inches to 1 yard 2 feet 8 inches, for there are 12 inches to a foot, and 3 feet to a yard.

You do the former if you care to; I'll do the latter. First, 6 inches and 8 inches are 14 inches. Divide 14 by 12, getting 1 and a remainder of 2, so you put down 2 and carry 1. As for the feet, $2+2+1=5$. Divide 5 by 3 and get 1 and a remainder of 2, put down 2 and carry 1. In the yards, you have $3+1+1=5$. Your answer, then, is 5 yards 2 feet 2 inches.

Now why on Earth should our unit ratios vary all over the lot, when our number system is so firmly based on 10? There are many reasons (valid in their time) for the use of odd ratios like 2, 3, 4, 8, 12, 16, and 20, but surely we are now advanced and sophisticated enough to use 10 as the exclusive (or nearly exclusive) ratio. If we could do so we could with such pleasure forget about compound addition—and compound subtraction, compound multiplication, compound division, too. (They also exist, of course.)

To be sure, there are times when nature makes the universal ten impossible. In measuring time, the day and the year have their lengths fixed for us by astronomical conditions and neither unit of time can be abandoned. Compound addition and the rest will have to be retained for such special cases, alas.

But who in blazes says we must measure things in firkins and pottles and Flemish ells? These are purely manmade measurements, and we must remember that measures were made for man and not man for measures.

It so happens that there is a system of measurement based exclusively on ten in this world. It is called the metric system and it is used all over the civilized world except for certain English-speaking nations such as the United States and Great Britain.

By not adopting the metric system, we waste our time for we gain nothing, not one thing, by learning our own measurements. The loss in time (which is expensive indeed) is balanced by not one thing I can imagine. (To be sure, it would be expensive to convert existing instruments and tools but it would have been nowhere nearly as expensive if we had done it a century ago, as we should have.)

There are those, of course, who object to violating our long-used cherished measures. They have given up cooms and chaldrons but imagine there is something about inches and feet and pints and quarts and pecks and bushels that is "simpler" or "more natural" than meters and liters.

There may even be people who find something dangerously foreign and radical (oh, for that vanished word of opprobrium, "Jacobin") in the metric system—yet it was the United States that led the way.

In 1786, thirteen years before the wicked French revolutionaries designed the metric system, Thomas Jefferson (a notorious "Jacobin," according to the Federalists, at least) saw a suggestion of his adopted by the infant United States. The nation established a decimal currency.

What we had been using was British currency, and that is a fearsome and wonderful thing. Just to point out how preposterous it is, let me say that the British people who, over the centuries, have, with monumental patience, taught themselves to endure anything at all provided it was "traditional"—are now sick and tired of their currency and are debating converting it to the decimal system. (They can't agree on the exact details of the change.)

But consider the British currency as it has been. To begin with, 4 farthings make 1 penny; 12 pennies make 1 shilling, and 20 shillings make 1 pound. In addition, there is a virtual farrago of terms, if not always actual coins, such as ha'pennies and thruppences and sixpences and crowns and half-crowns and florins and guineas and heaven knows what other devices with which to cripple the mental development of the British schoolchild and line the pockets of British tradesmen whenever tourists come to call and attempt to cope with the currency.

Needless to say, Pike gives careful instruction on how to manipulate pounds, shillings, and pence—and very special instructions they are. Try dividing 5 pounds, 13 shillings, 7 pence by 3. Quick now!

In the United States, the money system, as originally established, is as follows: 10 mills make 1 cent; 10 cents make 1 dime; 10 dimes make 1 dollar; 10 dollars make 1 eagle. Actually, modern Americans, in their calculations, stick to dollars and cents only.

The result? American money can be expressed in decimal form and can be treated as can any other decimals. An American child who has learned decimals need only be taught to recognize the dollar sign and he is all set. In the time that he does, a British child has barely mastered the fact that thruppence ha'penny equals 14 farthings.

What a pity that when, thirteen years later, in 1799, the metric system came into being, our original anti-British, pro-French feelings had not lasted just long enough to allow us to adopt it. Had we done so, we would have been as happy to forget our foolish pecks and ounces, as we are now happy to have forgotten our pence and shillings. (After all, would you like to go back to British currency in preference to our own?)

What I would like to see is one form of money do for all the world. Everywhere. Why not?

I appreciate the fact that I may be accused because of this of wanting to pour humanity into a mold, and of being a conformist. Of course, I am not a conformist (heavens!). I have no objection to local customs and local dialects and local dietaries. In fact, I insist on them for I constitute a locality all by myself. I just don't want to keep provincialisms that were well enough in their time but that interfere with human well-being in a world which is now 90 minutes in circumference.

If you think provincialism is cute and gives humanity color and charm, let me quote to you once more from Pike.

"Federal Money" (dollars and cents) had been introduced eleven years before Pike's second edition, and he gives the exact wording of the law that established it and discusses it in detail—under the decimal system and not under compound addition.

Naturally, since other systems than the Federal were still in use, rules had to be formulated and given for converting (or "reducing") one system to another. Here is the list. I won't give you the actual rules, just the list of reductions that were necessary, exactly as he lists them:

I. To reduce New Hampshire, Massachusetts, Rhode Island, Connecticut, and Virginia currency:

1. To Federal Money
2. To New York and North Carolina currency
3. To Pennsylvania, New Jersey, Delaware, and Maryland currency
4. To South Carolina and Georgia currency
5. To English money
6. To Irish money
7. To Canada and Nova Scotia currency
8. To Livres Tournois (French money)
9. To Spanish milled dollars

II. To reduce Federal Money to New England and Virginia currency:

III. To reduce New Jersey, Pennsylvania, Delaware, and Maryland currency:

1. To New Hampshire, Massachusetts, Rhode Island, Connecticut, and Virginia currency
2. To New York and . . .

Oh, the heck with it. You get the idea.

Can anyone possibly be sorry that all that cute provincial flavor has vanished? Are you sorry that every time you travel out of state you don't have to throw yourself into fits of arithmetical discomfort whenever you want to make a purchase? Or into similar fits every time someone from another state invades yours and tries to dicker with you? What a pleasure to have forgotten all that.

Then tell me what's so wonderful about having fifty sets of marriage and divorce laws?

In 1752, Great Britain and her colonies (some two centuries later than Catholic Europe) abondoned the Julian calendar and adopted the astronomically more correct Gregorian calendar (see Chapter 1). Nearly half a century later, Pike was still giving rules for solving complex calendar-based problems for the Julian calendar as well as for the Gregorian. Isn't it nice to have forgotten the Julian calendar?

Wouldn't it be nice if we could forget most of calendrical complications by adopting a rational calendar that would tie the day of the month firmly to the day of the week and have a single three-month calendar serve as a perpetual one, repeating itself over and over every three months? There is a world calendar proposed which would do just this.

It would enable us to do a lot of useful forgetting.

I would like to see the English language come into worldwide use. Not necessarily as the only language or even as the major language. It would just be nice if everyone—whatever his own language was—could also speak English fluently. It would help in communications and perhaps, eventually, everyone would just choose to speak English.

That would save a lot of room for other things.

Why English? Well, for one thing more people speak English as either first or second language than any other language on Earth, so we have a head start. Secondly, far more science is reported in English than in any other language and it is communication in science that is critical today and will be even more critical tomorrow.

To be sure, we ought to make it as easy as possible for people to speak English, which means we should rationalize its spelling and grammar.

English, as it is spelled today, is almost a set of Chinese ideograms. No one can be sure how a word is pronounced by looking at the letters that make it up. How do you pronounce: rough, through, though, cough, hiccough, and lough; and why is it so terribly necessary to spell all those sounds with the mad letter combination "ough"?

It looks funny, perhaps, to spell the words ruff, throo, thoh, cawf, hiccup, and lokh; but we already write hiccup and it doesn't look funny. We spell colour, color, and centre, center, and shew, show and grey, gray. The result looks funny to a Britisher but we are used to it. We can get used to the rest, too, and save a lot of wear and tear on the brain. We would all become more intelligent, if intelligence is measured by proficiency at spelling, and we'll not have lost one thing.

And grammar? Who needs the eternal hair-splitting arguments about "shall" and "will" or "which" and "that"? The uselessness of it can be demonstrated by the fact that virtually no one gets it straight anyway. Aside from losing valuable time, blunting a child's reasoning faculties, and instilling him or her with a ravening dislike for the English language, what do you gain?

If there be some who think that such blurring of fine distinctions will ruin the language, I would like to point out that English, before the grammarians got hold of it, had managed to lose its gender and its declensions almost everywhere except among the pronouns. The fact that we have only one definite article (the) for all genders and cases and times instead of three, as in French (*le, la, les*) or six, as in German *(der, die, das, dem, den, des)* in no way blunts the English language, which remains an admirably flexible instrument. We cherish our follies only because we are used to them and not because they are not really follies.

We must make room for expanding knowledge, or at least make as much room as possible. Surely it is as important to forget the old and useless as it is to learn the new and important.

Forget it, I say, forget it more and more. *Forget it!*

But why am I getting so excited? No one is listening to a word I say.

12. NOTHING COUNTS

In the previous chapter, I spoke of a variety of things; among them, Roman numerals. These seem, even after five centuries of obsolescence, to exert a peculiar fascination over the inquiring mind.

It is my theory that the reason for this is that Roman numerals appeal to the ego. When one passes a cornerstone which says: "Erected MCMXVIII," it gives one a sensation of power to say, "Ay, yes, nineteen eighteen" to one's self. Whatever the reason, they are worth further discussion.

The notion of number and of counting, as well as the names of the smaller and more-often-used numbers, date back to prehistoric times and I don't believe that there is a tribe of human beings on Earth today, however primitive, that does not have some notion of number.

With the invention of writing (a step which marks the boundary line between "prehistoric" and "historic"), the next step had to be taken—numbers had to be written. One can, of course, easily devise written symbols for the words that represent particular numbers, as easily as for any other word. In English we can write the number of fingers on one hand as "five" and the number of digits on all four limbs as "twenty."

Early in the game, however, the kings' tax-collectors, chroniclers, and scribes saw that numbers had the peculiarity of being ordered. There was one set way of counting numbers and any number could be defined by counting up to it. Therefore why not make marks which need be counted up to the proper number.

Thus, if we let "one" be represented as ′ and "two" as ″, and "three" as ‴, we can then work out the number indicated by a given symbol without trouble. You can see, for instance, that the symbol ′′′′′′′′′′′′′′′′′′′′′′′ stands for "twenty-three." What's more, such a symbol is universal. Whatever language you count in, the symbol stands for the number "twenty-three" in whatever sound your particular language uses to represent it.

It gets hard to read too many marks in an unbroken row, so it is only natural to break it up into smaller groups. If we are used to counting on the fingers of one hand, it seems natural to break up the marks into groups of five. "Twenty-three" then becomes ′′′′′ ′′′′′ ′′′′′ ′′′′′ ‴. If we are more sophisticated and use both hands in counting, we would write it ′′′′′′′′′′ ′′′′′′′′′′ ‴. If we go barefoot and use our toes, too, we might break numbers into twenties.

All three methods of breaking up number symbols into more easily handled groups have left their mark on the various number systems of mankind, but the favorite was division into ten. Twenty symbols in one group are, on the whole, too many for easy grasping, while five symbols in one group produce too many groups as numbers grow larger. Division into ten is the happy compromise.

It seems a natural thought to go on to indicate groups of ten by a separate mark. There is no reason to insist on writing out a group of ten as ′′′′′′′′′′ every time, when a separate mark, let us say -, can be used for the purpose. In that case "twenty-three" could be written as -- ‴.

Once you've started this way, the next steps are clear. By the time you have ten groups of ten (a hundred), you can introduce another symbol, for instance +. Ten hundreds, or a thousand, can become = and so on. In that case, the number "four thousand six hundred seventy-five" can be written = = = = + + + + + + ------- ′′′′′.

To make such a set of symbols more easily graspable, we can take advantage of the ability of the eye to form a pattern. (You know how you can tell the numbers displayed by a pack of cards or a pair of dice by the pattern itself.) We could therefore write "four thousand six hundred seventy-five" as

== + + + — '''
-
== + + + — ''.

And, as a matter of fact, the ancient Babylonians used just this system of writing numbers, but they used cuneiform wedges to express it.

The Greeks, in the earlier stages of their development, used a system similar to that of the Babylonians, but in later times an alternate method grew popular. They made use of another ordered system—that of the letters of the alphabet.

It is natural to correlate the alphabet and the number system. We are taught both about the same time in childhood, and the two ordered systems of objects naturally tend to match up. The series "ay, bee, see, dee . . ." comes as glibly as "one, two, three, four . . ." and there is no difficulty in substituting one for the other.

If we use undifferentiated symbols such as ''''''' for "seven," all the components of the symbol are identical and all must be included without exception if the symbol is to mean "seven" and nothing else. On the other hand, if "ABCDEFG" stands for "seven" (count the letters and see) then, since each symbol is different, only the last need be written. You can't confuse the fact that G is the seventh letter of the alphabet and therefore stands for "seven." In this way, a one-component symbol does the work of a seven-component symbol. Furthermore, '''''' (six) looks very much like ''''''' (seven); whereas F (six) looks nothing at all like G (seven).

The Greeks used their own alphabet, of course, but let's use our own alphabet here for the complete demonstration:

A = one, B = two, C = three, D = four, E = five, F = six, G = seven, H = eight, I = nine, and J = ten.

We could let the letter K go on to equal "eleven," but at that rate our alphabet will only help us up through "twenty-six." The Greeks had a better system. The Babylonian notion of groups of ten had left its mark. If J = ten, then J equals not only ten objects but also one group of tens. Why not, then, continue the next letters as numbering groups of tens?

In other words J = ten, K = twenty, L = thirty, M = forty, N = fifty, O = sixty, P = seventy, Q = eighty, R = ninety. Then we can go on to number groups of hundreds: S = one hundred, T = two hundred, U = three hundred, V = four hundred, W = five hundred, X = six hundred, Y = seven hundred, Z = eight hundred. It would be convenient to go on to nine hundred, but we have run out of letters. However, in old-fashioned alphabets the ampersand (&) was sometimes placed at the end of the alphabet, so we can say that & = nine hundred.

The first nine letters, in other words, represent the units from one to nine, the second nine letters represent the tens groups from one to nine, the third nine letters represent the hundreds groups from one to nine. (The Greek alphabet, in classic times, had only twenty-four letters where twenty-seven are needed, so the Greeks made use of three archaic letters to fill out the list.)

This system possesses its advantages and disadvantages over the Babylonian system. One advantage is that any number under a thousand can be given in three symbols. For instance, by the system I have just set up with our alphabet, six hundred seventy-five is XPE, while eight hundred sixteen is ZJF.

One disadvantage of the Greek system, however, is that the significance of twenty-seven different sumbols must be carefully memorized for the use of numbers to a thousand, whereas in the Babylonian system only three different symbols must be memorized.

Furthermore, the Greek system comes to a natural end when the letters of the alphabet are used up. Nine hundred ninety-nine (&RI) is the largest number that can be written without introducing special markings to indicate that a particular symbol indicates groups of thousands, tens of thousands, and so on. I will get back to this.

A rather subtle disadvantage of the Greek system was that the same symbols were used for numbers and words so that the mind could be easily distracted. For instance, the Jews of Graeco-Roman times adopted the Greek system of representing numbers but, of course, used the Hebrew alphabet—and promptly ran into a difficulty. The number "fifteen" would naturally be written as "ten-five." In the Hebrew alphabet, however, "ten-five" represents a short version of the ineffable name of the Lord, and the Jews, uneasy at the sacrilege, allowed "fifteen" to be represented as "nine-six" instead.

Worse yet, words in the Greek-Hebrew system look like numbers. For instance, to use our own alphabet, WRA is "five hundred ninety-one." In the alphabet system it doesn't usually matter in which order we place the symbols though, as we shall see, this came to be untrue for the Roman numerals, which are alphabetic, and WAR also means "five hundred ninety-one." (After all, we can say "five hundred one-and-ninety" if we wish.) Consequently, it is easy to believe that there is something warlike, martial, and of ominous import in the number "five hundred ninety-one."

The Jews, poring over every syllable of the Bible in their effort to copy the word of the Lord with the exactness that reverence required, saw numbers in all the words, and in New Testament times a whole system of mysticism arose over the numerical interrelationships within the Bible. This was the nearest the Jews came to mathematics, and they called this numbering of words *gematria*, which is a distortion of the Greek *geometria*. We now call it "numerology."

Some poor souls, even today, assign numbers to the different letters and decide which names are lucky and which unlucky, and which boy should marry which girl and so on. It is one of the more laughable pseudo-sciences.

In one case, a piece of gematria had repercussions in

later history. This bit of gematria is to be found in "The Revelation of St. John the Divine," the last book of the New Testament—a book which is written in a mystical fashion that defies literal understanding. The reason for the lack of clarity seems quite clear to me. The author of Revelation was denouncing the Roman government and was laying himself open to a charge of treason and to subsequent crucifixion if he made his words too clear. Consequently, he made an effort to write in such a way as to be perfectly clear to his "in-group" audience, while remaining completely meaningless to the Roman authorities.

In the thirteenth chapter he speaks of beasts of diabolical powers, and in the eighteenth verse he says, "Here is wisdom. Let him that hath understanding count the number of the beast: for it is the number of a man; and his number is Six hundred three-score and six."

Clearly, this is designed not to give the pseudo-science of gematria holy sanction, but merely to serve as a guide to the actual person meant by the obscure imagery of the chapter. Revelation, as nearly as is known, was written only a few decades after the first great persecution of Christians under Nero. If Nero's name ("Neron Caesar") is written in Hebrew characters the sum of the numbers represented by the individual letters does indeed come out to be six hundred sixty-six, "the number of the beast."

Of course, other interpretations are possible. In fact, if Revelation is taken as having significance for all time as well as for the particular time in which it was written, it may also refer to some anti-Christ of the future. For this reason, generation after generation, people have made attempts to show that, by the appropriate jugglings of the spelling of a name in an appropriate language, and by the appropriate assignment of numbers to letters, some particular personal enemy could be made to possess the number of the beast.

If the Christians could apply it to Nero, the Jews themselves might easily have applied it in the next century to Hadrian, if they had wished. Five centuries later it could be (and was) applied to Mohammed. At the time of the Reformation, Catholics calculated Martin Luther's name and found it to be the number of the beast, and Protestants returned the compliment by making the same discovery in the case of several popes.

Later still, when religious rivalries were replaced by nationalistic ones, Napoleon Bonaparte and William II were appropriately worked out. What's more, a few minutes' work with my own system of alphabet-numbers shows me that "Herr Adollf Hitler" has the number of the beast. (I need that extra "l" to make it work.)

The Roman system of number symbols had similarities to both the Greek and Babylonian systems. Like the Greeks, the Romans used letters of the alphabet. However, they did not use them in order, but used just a few letters which they repeated as often as necessary—as in the Babylonian system. Unlike the Babylonians, the Romans did not invent a new symbol for every tenfold increase of number, but (more primitively) used new symbols for fivefold increases as well.

Thus, to begin with, the symbol for "one" is I, and "two," "three," and four,' can be written II, III, and IIII.

The symbol for five, then, is *not* IIIII, but V. People have amused themselves no end trying to work out the reasons for the particular letters chosen as symbols, but there are no explanations that are universally accepted. However, it is pleasant to think that I represents the upheld finger and that V might symbolize the hand itself with all five fingers—one branch of the V would be the outheld thumb, the other, the remaining fingers. For "six" "seven," "eight," and "nine," we would then have VI, VII, VIII, and VIIII.

For "ten" we would then have X, which (some people think) represents both hands held wrist to wrist. "Twenty-three" would be XXIII, "forty-eight" would be XXXXVIII, and so on.

The symbol for "fifty" is L, for "one hundred" is C, for "five hundred" is D, and for "one thousand" is M. The C and M are easy to understand, for C is the first letter of *centum* (meaning "one hundred") and M is the first letter for *mille* (one thousand).

For that very reason, however, those symbols are suspicious. As initials they may have come to oust the original less-meaningful symbols for those numbers. For instance, an alternative symbol for "thousand" looks something like this (I). Half of a thousand or "five hundred" is the right half of the symbol, or (I), and this may have been converted into D. As for the L which stands for "fifty," I don't know why it is used.

Now, then, we can write nineteen sixty-four, in Roman numerals, as follows: MDCCCCLXIIII.

One advantage of writing numbers according to this system is that it doesn't matter in which order the numbers are written. If I decided to write nineteen sixty-four as CDCLIIMXCICI, it would still represent nineteen sixty-four if I add up the number values of each letter. However, it is not likely that anyone would ever scramble the letters in this fashion. If the letters were written in strict order of decreasing value, as I did the first time, it would then be much simpler to add the values of the letters. And, in fact, this order of decreasing value is (except for special cases) always used.

Once the order of writing the letters in Roman numerals is made an established convention, one can make use of deviations from that set order if it will help simplify matters. For instance, suppose we decide that when a symbol of smaller value *follows* one of larger value, the two are added; while if the symbol of smaller value *precedes* one of larger value, the first is subtracted from the second. Thus VI is "five" plus "one" or "six," while IV is "five" minus "one" or "four." (One might even say that IIV is "three," but it is conventional to subtract no more than one sumbol.) In the same way LX is "sixty" while XL is "forty"; CX is "one hundred ten," while XC is "ninety"; MC is "one thousand one hundred," while CM is "nine hundred."

The value of this "subtractive principle" is that two symbols can do the work of five. Why write VIIII if you can write IX; or DCCCC if you can write CM? The year nineteen sixty-four, instead of being written MDCCCCLXIIII (twelve symbols), can be written

MCMLXIV (seven symbols). On the other hand, once you make the order of writing letters significant, you can no longer scramble them even if you wanted to. For instance, if MMCLXIV is scrambled to MMCLXVI it becomes "two thousand one hundred sixty-six."

The subtractive principle was used on and off in ancient times but was not regularly adopted until the Middle Ages. One interesting theory for the delay involves the simplest use of the principle—that of IV ("four"). These are the first letters of IVPITER, the chief of the Roman gods, and the Romans may have had a delicacy about writing even the beginning of the name. Even today, on clockfaces bearing Roman numerals, "four" is represented as IIII and never as IV. This is not because the clockface does not accept the subtractive principle, for "nine" is represented as IX and never as VIIII.

With the symbols already given, we can go up to the number "four thousand nine hundred ninety-nine" in Roman numerals. This would be MMMMDCCCCLXXXXVIIII or, if the substractive principle is used, MMMMCMXCIX. You might suppose that "five thousand" (the next number) could be written MMMMM, but this is not quite right. Strictly speaking, the Roman system never requires a symbol to be repeated more than four times. A new symbol is always invented to prevent that: IIIII=V; XXXXX=L; and CCCCC=D. Well, then, what is MMMMM?

No letter was decided upon for "five thousand." In ancient times there was little need in ordinary life for numbers that high. And if scholars and tax collectors had occasion for larger numbers, their systems did not percolate down to the common man.

One method of penetrating to "five thousand" and beyond is to use a bar to represent thousands. Thus, $\overline{\text{V}}$ would represent not "five" but "five thousand." And sixty-seven thousand four hundred eighty-two would be $\overline{\text{LXVII}}$CDLXXXII.

But another method of writing large numbers harks back to the primitive symbol (I) for "thousand." By adding to the curved lines we can increase the number of ratios of ten. Thus "ten thousand" would be ((I)), and "one hundred thousand" would be (((I))). Then just as "five hundred" was I) or D, "five thousand" would be I)) and "fifty thousand" would be I))).

Just as the Romans made special marks to indicate thousands, so did the Greeks. What's more, the Greeks made special marks for ten thousands and for millions (or at least some of the Greek writers did). That the Romans didn't carry this to the logical extreme is no surprise. The Romans prided themselves on being non-intellectual. That the Greeks missed it also, however, will never cease to astonish me.

Suppose that instead of making special marks for large numbers only, one were to make special marks for every type of group from the units on. If we stick to the system I introduced at the start of the chapter—that is, the one in which ′ stands for units, - for tens, + for hundreds, and = for thousands—then we could get by with but one set of nine symbols. We could write every number with a little heading, marking off the type of groups = + −′. Then for "two thousand five hundred eighty-one" we could get by with only the letters from A to I and write it $\overset{=+-'}{\text{BEHA}}$. What's more, for "five thousand five hundred fifty-five" we could write $\overset{=+-'}{\text{EEEE}}$. There would be no confusion with all the E's, since the symbol above each E would indicate that one was a "five," another a "fifty," another a "five hundred," and another a "five thousand." By using additional symbols for ten thousands, hundred thousands, millions, and so on, any number, however large, could be written in this same fashion.

Yet it is not surprising that this would not be popular. Even if a Greek had thought of it he would have been repelled by the necessity of writing those tiny symbols. In an age of hand-copying, additional symbols meant additional labor and scribes would resent that furiously.

Of course, one might easily decide that the symbols weren't necessary. The Groups, one could agree, could always be written right to left in increasing values. The units would be at the right end, the tens next on the left, the hundreds next, and so on. In that case, BEHA would be "two thousand five hundred eighty-one" and EEEE would be "five thousand five hundred fifty-five" even without the little symbols on top.

Here, though, a difficulty would creep in. What if there were no groups of ten, or perhaps no units, in a particular number? Consider the number "ten" or the number "one hundred and one." The former is made up of one group of ten and no units, while the latter is made up of one group of hundreds, no groups of tens, and one unit. Using symbols over the columns, the numbers could be written
—′ +—′
A and AA, but now you would not dare leave out the little symbols. If you did, how could you differentiate A meaning "ten" from A meaning "one" or AA meaning "one hundred and one" from AA meaning "eleven" or AA meaning "one hundred and ten"?

You might try to leave a gap so as to indicate "one hundred and one" by A A. But then, in an age of hand-copying, how quickly would that become AA, or, for that matter, how quickly might AA become A A? Then, too, how would you indicate a gap at the end of a symbol? No, even if the Greeks thought of this system, they must obviously have come to the conclusion that the existence of gaps in numbers made this attempted simplification impractical. They decided it was safer to let J stand for "ten" and SA for "one hundred and one" and to Hades with little symbols.

What no Greek ever thought of—not even Archimedes himself—was that it wasn't absolutely necessary to work with gaps. One could fill the gap with a symbol by letting one stand for nothing—for "no groups." Suppose we use $ as such a symbol. Then, if "one hundred and one" is made up of one group of hundreds, no groups of tens, and
+—′
one unit, it can be written A$A. If we do that sort of thing, all gaps are eliminated and we don't need the little symbols on top. "One" becomes A, "ten" becomes A$, "one hundred" becomes A$$, "one hundred and one" becomes A$A, "one hundred and ten" becomes AA$, and so on.

Any number, however large, can be written with the use of exactly nine letters plus a symbol for nothing.

Surely this is the simplest thing in the world—after you think of it.

Yet it took men about five thousand years, counting from the beginning of number symbols, to think of a symbol for nothing. The man who succeeded (one of the most creative and original thinkers in history) is unknown. We know only that he was some Hindu who lived no later than the ninth century.

The Hindus called the symbol *sunya,* meaning "empty." This symbol for nothing was picked up by the Arabs, who termed it *sifr,* which in their language meant "empty." This has been distorted into our own words "cipher" and, by way of *zefirum,* into "zero."

Very slowly, the new system of numerals (called "Arabic numerals" because the Europeans learned of them from the Arabs) reached the West and replaced the Roman system.

Because the Arabic numerals came from lands which did not use the Roman alphabet, the shape of the numerals was nothing like the letters of the Roman alphabet and this was good, too. It removed word-number confusion and reduced gematria from the everyday occupation of anyone who could read, to a burdensome folly that only a few would wish to bother with.

The Arabic numerals as now used by us are, of course, 1, 2, 3, 4, 5, 6, 7, 8, 9, and the all-important 0. Such is our reliance on these numerals (which are internationally accepted) that we are not even aware of the extent to which we rely on them. For instance, if this chapter has seemed vaguely queer to you, perhaps it was because I had deliberately refrained from using Arabic numerals all through.

We all know the great simplicity Arabic numerals have lent to arithmetical computation. The unnecessary load they took off the human mind, all because of the presence of the zero, is simply incalculable. Nor has this fact gone unnoticed in the English language. The importance of the zero is reflected in the fact that when we work out an arithmetical computation we are (to use a term now slightly old-fashioned) "ciphering." And when we work out some code, we are "deciphering" it.

So if you look once more at the title of this chapter, you will see that I am not being cynical. I mean it literally. Nothing counts! The symbol for nothing makes all the difference in the world.

13. *C* FOR *CELERITAS*

If ever an equation has come into its own it is Einstein's $e = mc^2$. Everyone can rattle it off now, from the highest to the lowest; from the rarefied intellectual height of the science-fiction reader, through nuclear physicists, college students, newspaper reporters, housewives, busboys, all the way down to congressmen.

Rattling it off is not, of course, the same as understanding it; any more than a quick paternoster (from which, incidentally, the word "patter" is derived) is necessarily evidence of deep religious devotion.

So let's take a look at the equation. Each letter is the initial of a word representing the concept it stands for. Thus, *e* is the initial letter of "energy" and *m* of "mass." As for *c*, that is the speed of light in a vacuum, and if you ask why *c*, the answer is that it is the initial letter of *celeritas*, the Latin word meaning "speed."

This is not all, however. For any equation to have meaning in physics, there must be an understanding as to the units being used. It is meaningless to speak of a mass of 2.3, for instance. It is necessary to say 2.3 grams or 2.3 pounds or 2.3 tons; 2.3 alone is worthless.

Theoretically, one can choose whatever units are most convenient, but as a matter of convention, one system used in physics is to start with "grams" for mass, "centimeters" for distance, and "seconds" for time; and to build up, as far as possible, other units out of appropriate combinations of these three fundamental ones.

Therefore, the *m* in Einstein's equation is expressed in grams, abbreviated gm. The *c* represents a speed—that is, a distance traveled in a certain time. Using the fundamental units, this means the number of centimeters traveled in a certain number of seconds. The units of c are therefore centimeters per second, or cm/sec.

(Notice that the word "per" is represented by a fraction line. The reason for this is that to get a speed represented in lowest terms, that is, the number of centimeters traveled in *one* second, you must divide the number of centimeters traveled by the number of seconds of traveling. If you travel 24 centimeters in 8 seconds, your speed is 24 centimeters $\div$ 8 seconds, or 3 cm/sec.)

But, to get back to our subject, c occurs as its square in the equation. If you multiply c by c, you get c^2. It is, however, insufficient to multiply the numerical value of c by itself. You must also multiply the unit of c by itself.

A common example of this is in connection with measurements of area. If you have a tract of land that is 60 feet by 60 feet, the area is not 60×60, or 3600 feet. It is 60 feet $\times$ 60 feet, or 3600 square feet.

Similarly, in dealing with c^2, you must multiply cm/sec by cm/sec and end with the units cm^2/sec^2 (which can be read as centimeters squared per seconds squared).

The next question is: What is the unit to be used for *e?* Einstein's equation itself will tell us, if we remember to treat units as we treat any other algebraic symbols. Since $e = mc^2$, that means the unit of e can be obtained by multiplying the unit of m by the unit of c^2. Since the unit of m is gm and that of e^2 is em^2/sec^2, the unit of e is gm$\times$ cm^2/sec^2. In algebra we represent $a \times b$ as *ab;* consequently, we can run the multiplication sign out of the unit of e and make it simply gm cm^2/sec^2 (which is read "gram centimeter squared per second squared").

As it happens, this is fine, because long before Einstein worked out his equation it had been decided that the unit of energy on the gram-centimeter-second basis had to be gm cm^2/sec^2. I'll explain why this should be.

The unit of speed is, as I have said, cm/sec, but what happens when an object changes speed? Suppose that at a given instant, an object is traveling at 1 cm/sec, while a second later it is travelling at 2 cm/sec; and another second later it is traveling at 3 cm/sec. It is, in other words, "accelerating" (also from the Latin word *celeritas*).

In the case I've just cited, the acceleration is 1 centimeter per second every second, since each successive second it is going 1 centimeter per second faster. You might say that the acceleration is 1 cm/sec per second. Since we are letting the word "per" be represented by a fraction mark, this may be represented as 1 cm/sec/sec.

As I said before, we can treat the units by the same manipulations used for algebraic symbols. An expression like $a/b/b$ is equivalent to $a/b \div b$, which is in turn equivalent to $a/b \times 1/b$, which is in turn equivalent to a/b^2. By the same reasoning, 1 cm/sec/sec is equivalent to 1 cm/sec^2 and it is cm/sec^2 that is therefore the unit of acceleration.

A "force" is defined, in Newtonian physics, as something that will bring about an acceleration. By Newton's First Law of Motion any object in motion, left to itself, will travel at constant speed in a constant direction forever. A speed in a particular direction is referred to as a "velocity," so we might, more simply, say that an object in motion, left to itself, will travel at constant velocity forever. This velocity may well be zero, so that Newton's First Law also says that an object at rest, left to itself, will remain at rest forever.

As soon as a force, which may be gravitational, electromagnetic, mechanical, or anything, is applied, however, the velocity is changed. This means that its speed of travel or its direction of travel or both is changed.

The quantity of force applied to an object is measured by the amount of acceleration induced, and also by the mass of the object, since the force applied to a massive object produces less acceleration than the same force applied to a light object. (If you want to check this for yourself, kick a beach ball with all your might and watch it accelerate from rest to a good speed in a very short time. Next kick a cannon ball with all your might and observe—while hopping in agony—what an unimpressive acceleration you have imparted to it.)

To express this observed fact, one uses the expression: "Force equals mass times acceleration" or, to abbreviate, $f = ma$. Since the unit of mass is gm and the unit of acceleration is cm/sec^2, the unit of force is the product of the two or gm cm/sec^2.

Physicists grew tired of muttering "gram centimeter per second squared" every other minute, so they invented a single syllable to represent that phrase. The syllable is *dyne*, from the Greek *dynamis* meaning "power."

The multisyllabic expression and the monosyllable are equivalent: 1 dyne = 1 gm cm/sec^2. Dyne is just a breath-saver and can be defined as follows: A dyne is that amount of force which will impose upon a mass of one gram an acceleration of one centimeter per second squared.

See?

Next, there arises the problem of "work." Work as defined by the physicist is not what I do when I sit at the typewriter to write a chapter and slave my head to the very bone. To the physicist "work" is simply the motion of a body against a resisting force. To lift an object against the force of gravity is work; to pull a bar of iron away against the pull of a magnet is work; to drive a nail into the wood against the resistance of friction is work; and so on.

The amount of work done depends on the size of the resisting force and the distance moved against it. This can be expressed by saying: "Work equals force times distance," or, by abbreviation, $w = fd$.

The unit of distance is cm and the unit of force is dyne. Consequently, the unit of work is dyne cm. Again, physicists invented a monosyllable to express "dyne centimeters," and the new monosyllable is the ugly sound *erg*, from the Greek *ergon* meaning "work."

An erg is defined as the unit of work, and 1 erg is the amount of work performed by moving an object one cm against the resisting force of one dyne.

Lest you forget that this is all based on the gram-centimeter-second system, bring to mind the fact that a dyne is equivalent to a gm cm/sec^2. This means that the unit of work is cm times gm cm/sec^2 (distance times force), and this works out to gm cm^2/sec^2. In other words, 1 erg is the work done by imposing upon a mass of 1 gm an acceleration of 1 cm/sec^2 over a distance of 1 cm.

It was discovered a little over a century ago that work and energy are interconvertible, so that the units for one will serve as the units of the other. Consequently, the erg is also the unit of energy on the gram-centimeter-second basis.

Now shall we get back to Einstein's equation? There the units of e worked out to gm cm^2/sec^2, and that is equivalent to ergs. Those are the units we expect for energy, and it's no coincidence. If the equation had worked out to give any other units for energy, Einstein would have sharpened his pencil and started over again, knowing he had made a mistake.

Now we are ready to put numerical values into Einstein's equation. As far as m is concerned, we can suit ourselves and choose any convenient numerical quantity, the simplest choice being 1 gm.

In the case of c we have no option. The speed of light in a vacuum has a certain value and no other. In the units we have decided on, the best figure we have today is 29,-979,000,000 cm/sec. We wouldn't be far wrong in rounding this off to 30,000,000,000 cm/sec (a speed at which light can cover thirty billion centimeters—or three-quarters of the distance to the moon—in one second). Exponentially we can express this as 3×10^{10} cm/sec.

We have to square this to get the value of c^2, remembering to square both the number and the unit, and we end with 900,000,000,000,000,000,000 cm^2/sec^2 or 9×10^{20} cm^2/sec^2. The expression mc^2 (which is equal to e in Einstein's equation) thus becomes: 1 gm $\times$ 9 $\times$ 10^{20} cm^2/sec^2, which works out to 9×10^{20} gm cm^2/sec^2 or, if you prefer, 9×10^{20} ergs.

In other words, if 1 gram of matter were completely converted to energy, you would find yourself possessed of nine hundred quintillion ergs. And, on the other hand, if you wished to create 1 gram of matter out of pure energy (and could manage it with perfect efficiency), you would have to assure yourself, first, of a supply of nine hundred quintillion ergs.

This sounds impressive. Nine hundred quintillion ergs, wow!

But then, if you are cautious, you might stop and think: An erg is an unfamiliar unit. How large is it anyway?

After all, in Al Capp's Lower Slobbovia, the sum of a billion slobniks sounds like a lot until you find that the rate of exchange is ten billion slobniks to the dollar.

So—How large is an erg?

Well, it isn't large. As a matter of fact, it is quite a small unit. It is forced on physicists by the logic of the gram-centimeter-second system of units, but it ends in being so small a unit as to be scarcely useful. For instance, consider the task of lifting a pound weight one foot against gravity. That's not difficult and not much energy is expended. You could probably lift a hundred pounds one foot without completely incapacitating yourself. A professional strong man could do the same for a thousand pounds.

Nevertheless, the energy expended in lifting *one* pound one foot is equal to 13,558,200 ergs. Obviously, if any trifling bit of work is going to involve ergs in the tens of millions, we need other and larger units to keep the numerical values conveniently low.

For instance, there is an energy unit called a *joule,* which is equal to 10,000,000 ergs.

This unit is derived from the name of the British physicist James Prescott Joule, who inherited wealth and a brewery but spent his time in research. From 1840 to 1849 he ran a series of meticulous experiments which demonstrated conclusively the quantitative interconversion of heat and work and brought physics an understanding of the law of conservation of energy. However, it was the German scientist Hermann Ludwig Ferdinand von Helmholtz who first put the law into actual words in a paper presented in 1847, so that he consequently gets formal credit for the discovery.

(The word "joule," by the way is most commonly pronounced "jowl," although Joule himself probably pronounced his name "jool." In any case, I have heard over-precise people pronounce the word "zhool" under the impression that it is a French word, which it isn't. These are the same people who pronounce "centigrade" and "centrifuge," with a strong nasal twang as "sontigrade" and "sontrifuge" under the impression that these, too, are French words. Actually, they are Latin and no pseudo-French pronunciation is required. There is some justification for pronouncing "centimeter" as "sontimeter," since that is a French word to begin with, but in that case one should either stick to English or go French all the way and pronounce it "sontimettre," with a light accent on the third syllable.)

Anyway, notice the usefulness of the joule in everyday affairs. Lifting a pound mass a distance of one foot against gravity requires energy to the amount, roughly, of 1.36 joules—a nice, convenient figure.

Meanwhile, physicists who were studying heat had invented a unit that would be convenient for their purposes. This was the "calorie" (from the Latin word *calor* meaning "heat"). It can be abbreviated as cal. A calorie is the amount of heat required to raise the temperature of 1 gram of water from 14.5° C. to 15.5° C. (The amount of heat necessary to raise a gram of water one Celsius degree varies slightly for different temperatures, which is why one must carefully specify the 14.5 to 15.5 business.)

Once it was demonstrated that all other forms of energy and all forms of work can be quantitatively converted to heat, it could be seen that any unit that was suitable for heat would be suitable for any other kind of energy or work.

By actual measurement it was found (by Joule) that 4.185 joules of energy or work could be converted into precisely 1 calorie of heat. Therefore, we can say that 1 cal equals 4.185 joules equals 41,850,000 ergs.

Although the calorie, as defined above, is suitable for physicists, it is a little too small for chemists. Chemical reactions usually release or absorb heat in quantities that, under the conventions used for chemical calculations, result in numbers that are too large. For instance, 1 gram of carbohydrate burned to carbon dioxide and water (either in a furnace or the human body, it doesn't matter) liberates roughly 4000 calories. A gram of fat would, on burning, liberate roughly 9000 calories. Then again, a human being, doing the kind of work I do, would use up about 2,500,000 calories per day.

The figures would be more convenient if a larger unit were used, and for that purpose a larger calorie was invented, one that would represent the amount of heat required to raise the temperature of 1000 grams (1 kilogram) of water from 14.5° C. to 15.5° C. You see, I suppose, that this larger calorie is a thousand times as great as the smaller one. However, because both units are called "calorie," no end of confusion has resulted.

Sometimes the two have been distinguished as "small calorie" and "large calorie"; or "gram-calorie" and "kilogram-calorie"; or even "calorie" and "Calorie." (The last alternative is a particularly stupid one, since in speech—and scientists must occasionally speak—there is no way of distinguishing a *C* and a *c* by pronunciation alone.)

My idea of the most sensible way of handling the matter is this: In the metric system, a kilogram equals 1000 grams; a kilometer equals 1000 meters, and so on. Let's call the large calorie a kilocalorie (abbreviated kcal) and set it equal to 1000 calories.

In summary, then, we can say that 1 kcal equals 1000 cal or 4185 joules or 41,850,000,000 ergs.

Another type of energy unit arose in a roundabout way, via the concept of "power." Power is the rate at which work is done. A machine might lift a ton of mass one foot against gravity in one minute or in one hour. In each case the energy consumed in the process is the same, but it takes a more powerful heave to lift that ton in one minute than in one hour.

To raise one pound of mass one foot against gravity takes one *foot-pound* (abbreviated 1 ft-lb) of energy. To expand that energy in one second is to deliver 1 foot-pound per second (1 ft-lb/sec) and the ft-lb/sec is therefore a permissible unit of power.

The first man to make a serious effort to measure power accurately was James Watt (1736-1819). He compared the power of the steam engine he had devised with the power delivered by a horse, thus measuring his machine's rate of delivering energy in *horsepower* (or hp). In doing so, he first measured the power of a horse in ft-lb/sec and decided that 1 hp equals 550 ft-lb/sec, a conversion figure which is now standard and official.

The use of foot-pounds per second and horsepower is perfectly legitimate and, in fact, automobile and airplane engines have their power rated in horsepower. The trouble with these units, however, is that they don't tie in easily with the gram-centimeter-second system. A foot-pound is 1.355282 joules and a horsepower is 10.688 kilocalories per minute. These are inconvenient numbers to deal with.

The ideal gram-centimeter-second unit of power would be ergs per second (erg/sec). However, since the erg is such a small unit, it is more convenient to deal with joules per second (joule/sec). And since 1 joule is equal to 10,000,000 ergs, 1 joule/sec equals 10,000,000 erg/sec, or 10,000,000 gm cm^2/sec^3.

Now we need a monosyllable to express the unit joule/sec, and what better monosyllable than the monosyllabic name of the gentleman who first tried to measure power. So 1 joule/sec was set equal to 1 *watt*. The watt may be defined as representing the delivery of 1 joule of energy per second.

Now if power is multiplied by time, you are back to energy. For instance, if 1 watt is multipled by 1 second, you have 1 *watt-sec*. Since 1 watt equals 1 joule/sec, 1 watt-sec equals 1 joule/sec × sec, or 1 joule sec/sec. The secs cancel as you would expect in the ordinary algebraic manipulation to which units can be subjected, and you end with the statement that 1 watt-sec is equal to 1 joule and is, therefore, a unit of energy.

A larger unit of energy of this sort is the *kilowatt-hour* (or kw-hr). A kilowatt is equal to 100 watts and an hour is equal to 3600 seconds. Therefore a kw-hr is equal to 1000 × 3600 watt-sec, or to 3,600,000 joules, or to 36,-000,000,000,000 ergs.

Furthermore, since there are 4185 joules in a kilocalorie (kcal), 1 kw-hr is equal to 860 kcal or to 860,000 cal.

A human being who is living on 2500 kcal/day is delivering (in the form of heat, eventually) about 104 kcal/hr, which is equal to 0.120 kw hr/hr or 120 watts. Next time you're at a crowded cocktail party (or a crowded subway train or a crowded theater audience) on a hot evening in August, think of that as each additional person walks in. Each entrance is equivalent to turning on another one hundred twenty-watt electric bulb. It will make you feel a lot hotter and help you appreciate the new light of understanding that science brings.

But back to the subject. Now, you see, we have a variety of units into which we can translate the amount of energy resulting from the complete conversion of 1 gram of mass. That gram of mass will liberate:

	900,000,000,000,000,000,000	ergs,
or	90,000,000,000,000	joules,
or	21,500,000,000,000	calories,
or	21,500,000,000	kilocalories,
or	25,000,000	kilowatt-hours.

Which brings us to the conclusion that although the erg is indeed a tiny unit, nine hundred quintillion of them still mount up most impressively. Convert a mere one gram of mass into energy and use it with perfect efficiency and you can keep a thousand-watt electric light bulb running for 25,000,000 hours, which is equivalent to 2850 years, or the time from the days of Homer to the present.

How's that for solving the fuel problem?

We could work it the other way around, too. We might ask: How much mass need we convert to produce 1 kilowatt-hour of energy?

Well, if 1 gram of mass produces 25,000,000 kilowatt-hours of energy then 1 kilowatt-hour of energy is produced by 1/25,000,000 gram.

You can see that this sort of calculation is going to take us into small mass units indeed. Suppose we choose a unit smaller than the gram, say the *microgram*. This is equal to a millionth of a gram, i.e., 10^{-6} gram. We can then say that 1 kilowatt-hour of energy is produced by the conversion of 0.04 micrograms of mass.

Even the microgram is an inconveniently large unit of mass if we become interested in units of energy smaller than the kilowatt-hour. We could therefore speak of a *micromicrogram* (or, as it is now called, a *picogram*). This is a millionth of a millionth of a gram (10^{-12} gram) or a trillionth of a gram. Using that as a unit, we can say that:

1 kilowatt-hour	is	equivalent	to	40,000	picograms
1 kilocalorie	"	"	"	46.5	"
1 calorie	"	"	"	0.0465	"
1 joule	"	"	"	0.0195	"
1 erg	"	"	"	0.00000000195	"

To give you some idea of what this means, the mass of a typical human cell is about 1000 picograms. If, under conditions of dire emergency, the body possessed the ability to convert mass to energy, the conversion of the contents of 125 selected cells (which the body, with 50,000,000,000,000 cells or so, could well afford) would supply the body with 2500 kilocalories and keep it going for a full day.

The amount of mass which, upon conversion, yields 1 erg of energy (and the erg, after all, is the proper unit of energy in the gram-centimeter-second system) is an inconveniently small fraction even in terms of picograms.

We need units smaller still, so suppose we turn to the *picopicogram* (10^{-24} gram), which is a trillionth of a trillion of a gram, or a septillionth of a gram. Using the picopicogram, we find that it takes the conversion of 1950 picopicograms of mass to produce an erg of energy.

And the significance? Well, a single hydrogen atom has a mass of about 1.66 picopicograms. A uranium-235 atom has a mass of about 400 picopicograms. Consequently, an erg of energy is produced by the total conversion of 1200 hydrogen atoms or by 5 uranium-235 atoms.

In ordinary fission, only 1/1000 of the mass is converted to engery so it takes 5000 fissioning uranium atoms to produce 1 erg of energy. In hydrogen fusion, 1/100 of the mass is converted to energy, so it takes 120,000 fusing hydrogen atoms to produce 1 erg of energy.

And with that, we can let $e=mc^2$ rest for the nonce.

14. A PIECE OF THE ACTION

When my book *I, Robot* was reissued by the estimable gentlemen of Doubleday & Company, it was with a great deal of satisfaction that I noted certain reviewers (possessing obvious intelligence and good taste) beginning to refer to it as a "classic."

"Classic" is derived in exactly the same way, and has precisely the same meaning, as our own "first-class" and our colloquial "classy"; and any of these words represents my own opinion of *I, Robot,* too; except that (owing to my modesty) I would rather die than admit it. I mention it here only because I am speaking confidentially.

However, "classic" has a secondary meaning that displeases me. The word came into its own when the literary men of the Renaissance used it to refer to those works of the ancient Greeks and Romans on which they were modeling their own efforts. Consequently, "classic" has come to mean not only *good,* but also *old.*

Now *I, Robot* first appeared a number of years ago and some of the material in it was written ... Well, never mind. The point is that I have decided to feel a little hurt at being considered old enough to have written a classic, and therefore I will devote this chapter to the one field where "classic" is rather a term of insult.

Naturally, that field must be one where to be old is, almost automatically, to be wrong and incomplete. One may talk about Modern Art or Modern Literature or Modern Furniture and sneer as one speaks, comparing each, to their disadvantage, with the greater work of earlier ages.

When one speaks of Modern Science, however, one removes one's hat and places it reverently upon the breast.

In physics, particularly, this is the case. There is Modern Physics and there is (with an offhand, patronizing half-smile) Classical Physics. To put it into Modern Terminology, Modern Physics is in, man, in, and Classical Physics is like squaresville.

What's more, the division in physics is sharp. Everything after 1900 is Modern; everything before 1900 is Classical.

That looks arbitrary, I admit; a strictly parochial twentieth-century outlook. Oddly enough, though, it is perfectly legitimate. The year 1900 saw a major physical theory entered into the books and nothing has been quite the same since.

By now you have guessed that I am going to tell you about it.

The problem began with German physicist Gustav Robert Kirchhoff who, with Robert Wilhelm Bunsen (popularizer of the Bunsen burner), pioneered in the development of spectroscopy in 1859. Kirchhoff discovered that each element, when brought to incandescence, gave off certain characteristic frequencies of light; and that the vapor of the element, exposed to radiation from a source hotter than itself, absorbed just those frequencies it itself emitted when radiating. In short, a material will absorb those frequencies which, under other conditions, it will radiate; and will radiate those frequencies which, under other conditions, it will absorb.

But suppose that we consider a body which will absorb all frequencies of radiation that fall upon it—absorb them completely. It will then reflect none and will therefore appear absolutely black. It is a "black body." Kirchhoff pointed out that such a body, if heated to incandescence, would then necessarily have to radiate all frequencies of radiation. Radiation over a complete range in this manner would be "black-body radiation."

Of course, no body was absolutely black. In the 1890s, however, a German physicist named Wilhelm Wien thought of a rather interesting dodge to get around that. Suppose you had a furnace with a small opening. Any radiation that passes through the opening is either absorbed by the rough wall opposite or reflected. The reflected radiation strikes another wall and is again partially absorbed. What is reflected strikes another wall, and so on. Virtually none of the radiation survives to find its way out the small opening again. That small opening, then, absorbs the radiation and, in a manner of speaking, reflects none. It is a black body. If the furnace is heated, the radiation that streams out of that small opening should be black-body radiation and should, by Kirchhoff's reasoning, contain all frequencies.

Wien proceeded to study the characteristics of this black-body radiation. He found that at any temperature, a wide spread of frequencies was indeed included, but the spread was not an even one. There was a peak in the middle. Some intermediate frequency was radiated to a greater extent than other frequencies either higher or lower than that peak frequency. Moreover, as the temperature was increased, this peak was found to move toward the higher frequencies. If the absolute temperature were doubled, the frequency at the peak would also double.

But now the question arose: *Why* did black-body radiation distribute itself like this?

To see why the question was puzzling, let's consider infrared light, visible light, and ultraviolet light. The frequency range of infrared light, to begin with, is from one hundred billion (100,000,000,000) waves per second to four hundred trillion (400,000,000,000,000) waves per second. In order to make the numbers easier to handle let's divide by a hundred billion and number the frequency not in individual waves per second but in hundred-billion-wave packets per second. In that case the range of infrared would be from 1 to 4000.

Continuing to use this system, the range of visible light would be from 4000 to 8000; and the range of ultraviolet light would be from 8000 to 300,000.

Now it might be supposed that if a black body absorbed all radiation with equal ease, it ought to give off all radiation with equal ease. Whatever its temperature, the energy it had to radiate might be radiated at any frequency, the particular choice of frequency being purely random.

But suppose you were choosing numbers, *any* numbers with honest randomness, from 1 to 300,000. If you did this repeatedly, trillions of times, 1.3 per cent of your numbers would be less than 4000; another 1.3 per cent would be between 4000 and 8000, and 97.4 per cent would be between 8000 and 300,000.

This is like saying that a black body ought to radiate 1.3 per cent of its energy in the infrared, 1.3 per cent in visible light, and 97.4 per cent in the ultraviolet. If the temperature went up and it had more energy to radiate, it ought to radiate more at every frequency but the relative amounts in each range ought to be unchanged.

And this is only if we confine ourselves to nothing of still higher frequency than ultraviolet. If we include the x-ray frequencies, it would turn out that just about nothing should come off in the visible light at any temperature. Everything would be in ultraviolet and x-rays.

An English physicist, Lord Rayleigh (1842-1919), worked out an equation which showed exactly this. The radiation emitted by a black body increased steadily as one went up the frequencies. However, in actual practice, a frequency peak was reached after which, at higher frequencies still, the quantity of radiation decreased again. Rayleigh's equation was interesting but did not reflect reality.

Physicists referred to this prediction of the Rayleigh equation as the "Violet Catastrophe"—the fact that every body that had energy to radiate ought to radiate practically all of it in the ultraviolet and beyond.

Yet the whole point is that the Violet Catastrophe does not take place. A radiating body concentrated its radiation in the low frequencies. It radiated chiefly in the infrared at temperatures below, say, 1000° C., and radiated mainly in the visible region even at a temperature as high as 6000° C., the temperature of the solar surface.

Yet Rayleigh's equation was worked out according to

the very best principles available anywhere in physical theory—at the time. His work was an ornament of what we now call Classical Physics.

Wien himself worked out an equation which described the frequency distribution of black-body radiation in the high-frequency range, but he had no explanation for why it worked there, and besides it only worked for the high-frequency range, not for the low-frequency.

Black, black, black was the color of the physics mood all through the later 1890s.

But then arose in 1899 a champion, a German physicist, Max Karl Ernst Ludwig Planck. He reasoned as follows . . .

If beautiful equations worked out by impeccable reasoning from highly respected physical foundations do not describe the truth as we observe it, *then* either the reasoning or the physical foundations or both are wrong.

And *if* there is nothing wrong about the reasoning (and nothing wrong could be found in it), *then* the physical foundations had to be altered.

The physics of the day required that all frequencies of light be radiated with equal probability by a black body, and Planck therefore proposed that, on the contrary, they were *not* radiated with equal probability. Since the equal-probability assumption required that more and more light of higher and higher frequency be radiated, whereas the reverse was observed, Planck further proposed that the probability of radiation ought to decrease as frequency increased.

In that case, we would now have two effects. The first effect would be a tendency toward randomness which would favor high frequencies and increase radiation as frequency was increased. Second, there was the new Planck effect of decreasing probability of radiation as frequency went up. This would favor low frequencies and decrease radiation as frequency was increased.

In the low-frequency range the first effect is dominant, but in the high-frequency range the second effect increasingly overpowers the first. Therefore, in black-body radiation, as one goes up the frequencies, the amount of radiation first increases, reaches a peak, then decreases again—exactly as is observed.

Next, suppose the temperature is raised. The first effect can't be changed, for randomness is randomness. But suppose that as the temperature is raised, the probability of emitting high-frequency radiation increases. The second effect, then, is steadily weakened as the temperature goes up. In that case, the radiation continues to increase with increasing frequency for a longer and longer time before it is overtaken and repressed by the gradually weakening second effect. The peak radiation, consequently, moves into higher and higher frequencies as the temperature goes up—precisely as Wien had discovered.

On this basis, Planck was able to work out an equation that described black-body radiation very nicely both in the low-frequency and high-frequency range.

However, it is all very well to say that the higher the frequency the lower the probability of radiation, but *why?* There was nothing in the physics of the time to explain that, and Planck had to make up something new.

Suppose that energy did not flow continuously, as physicists had always assumed, but was given off in pieces. Suppose there were "energy atoms" and these increased in size as frequency went up. Suppose, still further, that light of a particular frequency could not be emitted unless enough energy had been accumulated to make up an "energy atom" of the size required by that frequency.

The higher the frequency the larger the "energy atom" and the smaller the probability of its accumulation at any given instant of time. Most of the energy would be lost as radiation of lower frequency, where the "energy atoms" were smaller and more easily accumulated. For that reason, an object at a temperature of 400° C. would radiate its heat in the infrared entirely. So few "energy atoms" of visible light size would be accumulated that no visible glow would be produced.

As temperature went up, more energy would be generally available and the probabilities of accumulating a high-frequency "energy atom" would increase. At 6000° C. most of the radiation would be in "energy atoms" of visible light, but the still larger "energy atoms" of ultraviolet would continue to be formed only to a minor extent.

But how big is an "energy atom"? How much energy does it contain? Since this "how much" is a key question, Planck, with admirable directness, named the "energy atom" a *quantum,* which is Latin for "how much?" The plural is *quanta.*

For Planck's equation for the distribution of black-body radiation to work, the size of the quantum had to be directly proportional to the frequency of the radiation. To express this mathematically, let us represent the size of the quantum, or the amount of energy it contains, by e (for energy). The frequency of radiation is invariably represented by physicists by means of the Greek letter nu (ν).

If energy (e) is proportional to frequency (ν), then e must be equal to ν multiplied by some constant. This constant, called *Planck's constant,* is invariably represented as h. The equation, giving the size of a quantum for a particular frequency of radiation, becomes:

$$e = h\nu \qquad \text{(Equation 1)}$$

It is this equation, presented to the world in 1900, which is the Continental Divide that separates Classical Physics from Modern Physics. In Classical Physics, energy was considered continuous; in Modern Physics it is considered to be composed of quanta. To put it another way, in Classical Physics the value of h is considered to be 0; in Modern Physics it is considered to be greater than 0.

It is as though there were a sudden change from considering motion as taking place in a smooth glide, to motion as taking place in a series of steps.

There would be no confusion if steps were long galumphing strides. It would be easy, in that case, to distinguish steps from a glide. But suppose one minced along in microscopic little tippy-steps, each taking a tiny fraction of a second. A careless glance could not distinguish that from a glide. Only a painstaking study would show that your head was bobbing slightly with each step. The smaller the steps, the harder to detect the difference from a glide.

In the same way, everything would depend on just how big individual quanta were; on how "grainy" energy was. The size of the quanta depends on the size of Planck's constant, so let's consider that for a while.

If we solve Equation 1 for *h*, we get:

$$h = e/\nu \qquad \text{(Equation 2)}$$

Energy is very frequently measured in ergs (see Chapter 13). Frequency is measured as "so many per second" and its units are therefore "reciprocal seconds" or "1/second."

We must treat the units of *h* as we treat *h* itself. We get *h* by dividing *e* by *v*; so we must get the units of *h* by dividing the units of *e* by the units of *v*. When we divide ergs by 1/second we are multiplying ergs by seconds, and we find the units of *h* to be "erg-seconds." A unit which is the result of multiplying energy by time is said, by physicists, to be one of "action." Therefore, Planck's constant is expressed in units of action.

Since the nature of the universe depends on the size of Planck's constant, we are all dependent on the size of the piece of action it represents. Planck, in other words, had sought and found *the* piece of the action. (I understand that others have been searching for a piece of the action ever since, but where's the point since Planck has found it?)

And what is the exact size of *h*? Planck found it had to be very small indeed. The best value, currently accepted, is: 0.0000000000000000000000000066256 erg-seconds or 6.6256×10^{-27} erg-seconds.

Now let's see if I can find a way of expressing just how small this is. The human body, on an average day, consumes and expends about 2500 kilocalories in maintaining itself and performing its tasks. One kilocalorie is equal to 1000 calories, so the daily supply is 2,500,000 calories.

One calorie, then, is a small quantity of energy from the human standpoint. It is 1/2,500,000 of your daily store. It is the amount of energy contained in 1/113,000 of an ounce of sugar, and so on.

Now imagine you are faced with a book weighing one pound and wish to lift it from the floor to the top of a bookcase three feet from the ground. The energy expended in lifting one pound through a distance of three feet against gravity is just about 1 calorie.

Suppose that Planck's constant were of the order of a calorie-second in size. The universe would be a very strange place indeed. If you tried to lift the book, you would have to wait until enough energy had been accumulated to make up the tremendously sized quanta made necessary by so large a piece of action. Then, once it was accumulated, the book would suddenly be three feet in the air.

But a calorie-second is equal to 41,850,000 erg-seconds, and since Planck's constant is such a minute fraction of one erg-second, a single calorie-second equals 6,385,400,000,000,000,000,000,000,000,000,000 Planck's constants, or 6.3854×10^{33} Planck's constants, or about six and a third decillion Planck's constants. However you slice it, a calorie-second is equal to a tremendous number of Planck's constants.

Consequently, in any action such as the lifting of a one-pound book, matters are carried through in so many trillions of trillions of steps, each one so tiny, that motion seems a continuous glide.

When Planck first introduced his "quantum theory" in 1900, it caused remarkably little stir, for the quanta seemed to be pulled out of midair. Even Planck himself was dubious—not over his equation describing the distribution of black-body radiation, to be sure, for that worked well; but about the quanta he had introduced to explain the equation.

Then came 1905, and in that year a 26-year-old theoretical physicist, Albert Einstein, published five separate scientific papers on three subjects, any one of which would have been enough to establish him as a first-magnitude star in the scientific heavens.

In two, he worked out the theoretical basis for "Brownian motion" and, incidentally, produced the machinery by which the actual size of atoms could be established for the first time. It was one of these papers that earned him his Ph.D.

In the third paper, he dealt with the "photoelectric effect" and showed that although Classical Physics could not explain it, Planck's quantum theory could.

This really startled physicists. Planck had invented quanta merely to account for black-body radiation, and here it turned out to explain the photoelectric effect, too, something entirely different. For quanta to strike in two different places like this, it seemed suddenly very reasonable to suppose that they (or something very like them) actually existed.

(Einstein's fourth and fifth papers set up a new view of the universe which we call "The Special Theory of Relativity." It is in these papers that he introduced his famous equation $e = mc^2$; see Chapter 13.

These papers on relativity, expanded into a "General Theory" in 1915, are the achievements for which Einstein is known to people outside the world of physics. Just the same, in 1921, when he was awarded the Nobel Prize for Physics, it was for his work on the photoelectric effect and *not* for his theory of relativity.)

The value of h is so incredibly small that in the ordinary world we can ignore it. The ordinary gross events of everyday life can be considered as though energy were a continuum. This is a good "first approximation."

However, as we deal with smaller and smaller energy changes, the quantum steps by which those changes must take place become larger and larger in comparison. Thus, a flight of stairs consisting of treads 1 millimeter high and 3 millimeters deep would seem merely a slightly roughened ramp to a six-foot man. To a man the size of an ant, however, the steps would seem respectable individual obstacles to be clambered over with difficulty. And to a man the size of a bacterium, they would be mountainous precipices.

In the same way, by the time we descend into the world within the atom the quantum step has become a gigantic thing. Atomic physics cannot, therefore, be described in Classical terms, not even as an approximation.

The first to realize this clearly was the Danish physicist Niels Bohr. In 1913 Bohr pointed out that if an electron absorbed energy, it had to absorb it a whole quantum at a time and that to an electron a quantum was a large piece of energy that forced it to change its relationship to the rest of the atom drastically and all at once.

Bohr pictured the electron as circling the atomic nucleus in a fixed orbit. When it absorbed a quantum of energy, it suddenly found itself in an orbit farther from the nucleus—there was no in-between, it was a one-step proposition.

Since only certain orbits were possible, according to Bohr's treatment of the subject, only quanta of certain size could be absorbed by the atom—only quanta large enough to raise an electron from one permissible orbit to another. When the electrons dropped back down the line of permissible orbits, they emitted radiations in quanta. They emitted just those frequencies which went along with the size of quanta they could emit in going from one orbit to another.

In this way, the science of spectroscopy was rationalized. Men understood a little more deeply why each element (consisting of one type of atom with one type of energy relationships among the electrons making up that type of atom) should radiate certain frequencies, and certain frequencies only, when incandescent. They also understood why a substance that could absorb certain frequencies should also emit those same frequencies under other circumstances.

In other words, Kirchhoff had started the whole problem and now it had come around full-circle to place his empirical discoveries on a rational basis.

Bohr's initial picture was oversimple; but he and other men gradually made it more complicated, and capable of explaining finer and finer points of observation. Finally, in 1926, the Austrian physicist Erwin Schrödinger worked out a mathematical treatment that was adequate to analyze the workings of the particles making up the interior of the atom according to the principles of the quantum theory. This was called "quantum mechanics," as opposed to the "classical mechanics" based on Newton's three laws of motion and it is quantum mechanics that is the foundation of Modern Physics.

15. WELCOME, STRANGER!

There are fashions in science as in everything else. Conduct an experiment that brings about an unusual success and before you can say, "There are a dozen imitations!" there are a dozen imitations!

Consider the element xenon (pronounced zee′non), discovered in 1898 by William Ramsay and Morris William Travers. Like other elements of the same type it was isolated from liquid air. The existence of these elements in air had remained unsuspected through over a century of ardent chemical analysis of the air, so when they finally dawned upon the chemical consciousness they were greeted as strange and unexpected newcomers. Indeed, the name, xenon, is the neutral form of the Greek word for "strange," so that xenon is "the strange one" in all literalness.

Xenon belongs to a group of elements commonly known as the "inert gases" (because they are chemically inert) or the "rare gases" (because they are rare), or "noble gases" because the standoffishness that results from chemical inertness seems to indicate a haughty sense of self-importance.

Xenon is the rarest of the stable inert gases and, as a matter of fact, is the rarest of all the stable elements on Earth. Xenon occurs only in the atmosphere, and there it makes up about 5.3 parts per million by weight. Since the atmosphere weighs about 5,500,000,000,000,000 (five and a half quadrillion) tons, this means that the planetary supply of xenon comes to just about 30,000,000,000 (thirty billion) tons. This seems ample, taken in full, but picking xenon atoms out of the overpoweringly more common constituents of the atmosphere is an arduous task and so xenon isn't a common substance and never will be.

What with one thing and another, then, xenon was not a popular substance in the chemical laboratories. Its chemical, physical, and nuclear properties were working out, but beyond that there seemed little worth doing with it. It remained the little strange one and received cold shoulders and frosty smiles.

Then, in 1962, an unusual experiment involving xenon was announced whereupon from all over the world broad smiles broke out across chemical countenances, and little xenon was led into the test tube with friendly solicitude. "Welcome, stranger!" was the cry everywhere, and now you can't open a chemical journal anywhere without finding several papers on xenon.

What happened?

If you expect a quick answer, you little know me. Let me take my customary route around Robin Hood's barn and begin by stating, first of all, that xenon is a gas.

Being a gas is a matter of accident. No substance is a gas intrinsically, but only insofar as temperature dictates. On Venus, water and ammonia are both gases. On Earth, ammonia is a gas, but water is not. On Titan, neither ammonia nor water are gases.

So I'll have to set up an arbitrary criterion to suit my present purpose. Let's say that any substance that remains a gas at —100° C. (—148° F.) is a Gas with a capital letter, and concentrate on those. This is a temperature that is never reached on Earth, even in an Antarctic winter of extraordinary severity, so that no Gas is ever anything but gaseous on Earth (except occasionally in chemical laboratories).

Now why is a Gas a Gas?

I can start by saying that every substance is made up of atoms, or of closely knit groups of atoms, said groups being called molecules. There are attractive forces between atoms or molecules which make them "sticky" and tend to hold them together. Heat, however, lends these atoms or molecules a certain kinetic energy (energy of motion) which tends to drive them apart, since each atom or molecule has its own idea of where it wants to go.*

The attractive forces among a given set of atoms or molecules are relatively constant, but the kinetic energy varies with the temperature. Therefore, if the temperature is raised high enough, any group of atoms or molecules will fly apart and the material becomes a gas. At temperatures over 6000° C. all known substances are gases.

Of course, there are only a few exceptional substances with interatomic or intermolecular forces so strong that it takes 6000° C. to overcome them. Some substances, on the other hand, have such weak intermolecular attractive forces that the warmth of a summer day supplies enough kinetic energy to convert them to gas (the common anesthetic, ether, is an example).

Still others have intermolecular attractive forces so much weaker still that there is enough heat at a temperature of —100° C. to keep them gases, and it is these that are the Gases I am talking about.

The intermolecular or interatomic forces arise out of the distribution of electrons within the atoms or molecules. The electrons are distributed among various "electron shells," according to a system we can accept without detailed explanation. For instance, the aluminum atom contains 13 electrons, which are distributed as follows: 2 in the innermost shell, 8 in the next shell, and 3 in the next shell. We can therefore signify the electron distribution in the aluminum atom as 2,8,3.

The most stable and symmetrical distribution of the electrons among the electron shells is that distribution in which the outermost shell holds either all the electrons it can hold, or 8 electrons—whichever is less. The innermost electron shell can hold only 2, the next can hold 8, and each of the rest can hold more than 8. Except for the situation where only the innermost shell contains electrons,

* No, I am not implying that atoms know what they are doing and have consciousness. This is just my teleological way of talking. Teleology is forbidden in scientific articles, but it so happens I enjoy sin.

then, the stable situation consists of 8 electrons in the outermost shell.

There are exactly six elements known in which this situation of maximum stability exists:

Element	*Symbol*	*Electron Distribution*	*Electron Total*
helium	He	2	2
neon	Ne	2,8	10
argon	Ar	2,8,8	18
krypton	Kr	2,8,18,8	36
xenon	Xe	2,8,18,18,8	54
radon	Rn	2,8,18,32,18,8	86

Other atoms without this fortunate electronic distribution are forced to attempt to achieve it by grabbing additional electrons, or getting rid of some they already possess, or sharing electrons. In so doing, they undergo chemical reactions. The atoms of the six elements listed above, however, need do nothing of this sort and are sufficient unto themselves. They have no need to shift electrons in any way and that means they take part in no chemical reactions and are inert. (At least, this is what I would have said prior to 1962.)

The atoms of the inert gas family listed above are so self-sufficient, in fact, that the atoms even ignore one another. There is little interatomic attraction, so that all are gases at room temperature and all but radon are Gases.

To be sure, there is *some* interatomic attraction (for no atoms or molecules exist among which there is no attraction at all). If one lowers the temperature sufficiently, a point is reached where the attractive forces become dominant over the disruptive effect of kinetic energy, and every single one of the inert gases will, eventually, become an inert liquid.

What about other elements? As I said, these have atoms with electron distributions of less than maximum stability and each has a tendency to alter that distribution in the direction of stability. For instance, the sodium atom (Na) has a distribution of 2,8,1. If it could get rid of the outermost electron, what would be left would have the stable 2,8 configuration of neon. Again, the chlorine atom (Cl) has a distribution of 2,8,7. If it could gain an electron, it would have the 2,8,8 distribution of argon.

Consequently, if a sodium atom encounters a chlorine atom, the transfer of an electron from the sodium atom to the chlorine atom satisfies both. However, the loss of a negatively charged electron leaves the sodium atom with a deficiency of negative charge or, which is the same thing, an excess of positive charge. It becomes a positively charged sodium ion (Na^+). The chlorine atom, on the other hand, gaining an electron, gains an excess of negative charge and becomes a negatively charged chloride ion* (Cl^-).

Opposite charges attract, so the sodium ion attracts all the chloride ions within reach and vice versa. These strong attractions cannot be overcome by the kinetic energy induced at ordinary temperatures, and so the ions hold together firmly enough for "sodium chloride" (common salt) to be a solid. It does not become a gas, in fact, until a temperature of 1413° C. is reached.

Next, consider the carbon atom (C). Its electron distribution is 2,4. If it lost 4 electrons, it would gain the 2 helium configuration; if it gained 4 electrons, it would gain the 2,8 neon configuration. Losing or gaining that many electrons is not easy, so the carbon atom shares electrons instead. It can, for instance, contribute one of its electrons to a "shared pool" of two electrons, a pool to which a neighboring carbon atom also contributes an electron. With its second electron it can form another shared pool with a second neighbor, and with its third and fourth, two more pools with two more neighbors. Each neighbor can set up additional pools with other neighbors. In this way, each carbon atom is surrounded by four other carbon atoms.

* The charged chlorine atom is called "chloride ion" and not "chlorine ion" as a convention of chemical nomenclature we might just as well accept with a weary sigh. Anyway, the "d" is not a typographical error.

These shared electrons fit into the outermost electron shells of each carbon atom that contributes. Each carbon atom has 4 electrons of its own in that outermost shell and 4 electrons contributed (one apiece) by four neighbors. Now, each carbon atom has the 2,8 configuration of neon, but only at the price of remaining close to its neighbors. The result is a strong interatomic attraction, even though electrical charge is not involved. Carbon is a solid and is not a gas until a temperature of 4200° C. is reached.

The atoms of metallic elements also stick together strongly, for similar reasons, so that tungsten, for instance, is not a gas until a temperature of 5900° C. is reached.

We cannot, then, expect to have a gas when atoms achieve stable electron distribution by transferring electrons in such a manner as to gain an electric charge; or by sharing electrons in so complicated a fashion that vast numbers of atoms stick together in one piece.

What we need is something intermediate. We need a situation where atoms achieve stability by sharing electrons (so that no electric charge arises) but where the total number of atoms involved in the sharing is very small so that only small molecules result. Within the molecules, attractive forces may be large, and the molecules may not be shaken apart without extreme temperature. The attractive forces between one molecule and its neighbor, however, may be small—and that will do.

Let's consider the hydrogen atom, for instance. It has but a single electron. Two hydrogen atoms can each contribute its single electron to form a shared pool. As long as they stay together, each can count both electrons in its outermost shell and each will have the stable helium configuration. Furthermore, neither hydrogen atom will have any electrons left to form pools with other neighbors, hence the molecule will end there. Hydrogen gas will consist of two-atom molecules (H_2).

The attractive force between the atoms in the molecule is large, and it takes temperatures of more than 2000° C. to shake even a small fraction of the hydrogen molecules into single atoms. There will, however, be only weak attractions among separate hydrogen molecules, each of which, under the new arrangement, will have reached a satisfactory pitch of self-sufficiency. Hydrogen, therefore, will be a Gas—one not made up of separate atoms as is the case with the inert gases, but of two-atom molecules.

Something similar will be true in the case of fluorine (electronic distribution 2,7), oxygen (2,6) and nitrogen (2,5). The fluorine atom can contribute an electron and form a shared pool of two electrons with a neighboring fluorine atom which also contributes an electron. Two oxygen atoms can contribute two electrons apiece to form a shared pool of four electrons, and two nitrogen atoms can contribute three electrons each and form a shared pool of six electrons.

In each case, the atoms will achieve the 2,8 distribution of neon at the cost of forming paired molecules. As a result, enough stability is achieved so that fluorine (F_2), oxygen (O_2), and nitrogen (N_2) are all Gases.

The oxygen atom can also form a shared pool of two electrons with each of two neighbors, and those two neighbors can form another shared pool of two electrons among themselves. The result is a combination of three oxygen atoms (O_3), each with a neon configuration. This combination, O_3, is called ozone, and it is a Gas too.

Oxygen, nitrogen, and fluorine can form mixed molecules, too. For instance, a nitrogen and an oxygen atom can combine to achieve the necessary stability for each. Nitrogen may also form shared pools of two electrons with each of three fluorine atoms, while oxygen may do so with each of two. The resulting compounds: nitrogen oxide (NO), nitrogen trifluoride. (NF_3), and oxygen difluoride (OF_2) are all Gases.

Atoms which, by themselves, will not form Gases may do so if conbined with either hydrogen, oxygen, nitrogen, or fluorine. For instance, two chlorine atoms (2,8,7, remember) will form a shared pool of two electrons so that both achieve the 2,8,8 argon configuration. Chlorine (Cl_2) is therefore a gas at room temperature—with intermolecular attractions, however, large enough to keep it from being a Gas. Yet if a chlorine atom forms a shared pool of two electrons with a fluorine atom, the result, chlorine fluoride (ClF), *is* a Gas.

The boron atom (2,3) can form a shared pool of two electrons with each of three fluorine atoms, and the carbon atom a shared pool of two electrons with each of four fluorine atoms. The resulting compounds, boron trifluoride (BF_3) and carbon tetrafluoride (CF_4), are Gases.

A carbon atom can form a shared pool of two electrons with each of four hydrogen atoms, or a shared pool of four electrons with an oxygen atom, and the resulting compounds, methane (CH_4) and carbon monoxide (CO), are gases. A two-carbon combination may set up a shared pool of two electrons with each of four hydrogen atoms (and a shared pool of four electrons with one another); a silicon atom may set up a shared pool of two electrons with each of four hydrogen atoms. The compounds, ethylene (C_2H_4) and silane (SiH_4), are Gases.

Altogether, then, I can list twenty Gases which fall into the following categories:

(1) Five elements made up of single atoms: helium, neon, argon, krypton, and xenon.

(2) Four elements made up of two-atom molecules: hydrogen, nitrogen, oxygen, and fluorine.

(3) One element form made up of three-atom molecules: ozone (of oxygen).

(4) Ten compounds, with molecules built up of two different elements, at least one of which falls into category (2).

The twenty Gases are listed in order of increasing boiling point in the accompanying table, and that boiling point is given in both the Celsius scale (°C.) and the Absolute scale (°K.).

The five inert gases on the list are scattered among the fifteen other Gases. To be sure, two of the three lowest-boiling Gases are helium and neon, but argon is seventh, krypton is tenth, and xenon is seventeenth. It would not be surprising if all the Gases, then, were as inert as the inert gases.

The Twenty Gases

Substance	*Formula*	*B.P. (C.°)*	*B.P. (K.°)*
Helium	He	−268.9	4.2
Hydrogen	H_2	−252.8	20.3
Neon	Ne	−245.9	27.2
Nitrogen	N_2	−195.8	77.3
Carbon monoxide	CO	−192	81
Fluorine	F_2	−188	85
Argon	Ar	−185.7	87.4
Oxygen	O_2	−183.0	90.1
Methane	CH_4	−161.5	111.6
Krypton	Kr	−152.9	120.2
Nitrogen oxide	NO	−151.8	121.3
Oxygen difluoride	OF_2	−144.8	128.3
Carbon tetrafluoride	CF_4	−128	145
Nitrogen trifluoride	NF_3	−120	153
Ozone	O_3	−111.9	161.2
Silane	SiH_4	−111.8	161.3
Xenon	Xe	−107.1	166.0
Ethylene	C_2H_4	−103.9	169.2
Boron trifluoride	BF_3	−101	172
Chlorine fluoride	ClF	−100.8	172.3

Perhaps they might be at that, if the smug, self-sufficient molecules that made them up were permanent, unbreakable affairs, but they are not. All the molecules can be broken down under certain conditions, and the free atoms (those of fluorine and oxygen particularly) are active indeed.

This does not show up in the Gases themselves. Suppose a fluorine molecule breaks up into two fluorine atoms, and these find themselves surrounded only by fluorine molecules? The only possible result is the re-formation of a fluorine molecule, and nothing much has happened. If, however, there are molecules other than fluorine present, a new molecular combination of greater stability than F_2 is possible (indeed, almost certain in the case of fluorine), and a chemical reaction results.

The fluorine molecule does have a tendency to break apart (to a very small extent) even at ordinary temperatures, and this is enough. The free fluorine atom will attack virtually anything non-fluorine in sight, and the heat of reaction will raise the temperature, which will bring about a more extensive split in fluorine molecules, and so on. The result is that molecular fluorine is the most chemically active of all the Gases (with chlorine fluoride almost on a par with it and ozone making a pretty good third).

The oxygen molecule is torn apart with greater difficulty and therefore remains intact (and inert) under conditions where fluorine will not. You may think that oxygen is an active element, but for the most part this is only true under elevated temperatures, where more energy is available to tear it apart. After all, we live in a sea of free oxygen without damage. Inanimate substances such as paper, wood, coal, and gasoline, all considered flammable, can be bathed by oxygen for indefinite periods without perceptible chemical reaction—until heated.

Of course, once heated, oxygen does become active and combines easily with other Gases such as hydrogen, carbon monoxide, and methane which, by that token, can't be considered particularly inert either.

The nitrogen molecule is torn apart with still more difficulty and, before the discovery of the inert gases, nitrogen was *the* inert gas *par excellence.* It and carbon tetrafluoride are the only Gases on the list, other than the inert gases themselves, that are respectably inert, but even they can be torn apart.

Life depends on the fact that certain bacteria can split the nitrogen molecule; and important industrial processes arise out of the fact that man has learned to do the same thing on a large scale. Once the nitrogen molecule is torn apart, the individual nitrogen atom is quite active, bounces around in all sorts of reactions and, in fact, is the fourth most common atom in living tissue and is essential to all its workings.

In the case of the inert gases, all is different. There are no molecules to pull apart. We are dealing with the self-sufficient atom itself, and there seemed little likelihood that combination with any other atom would produce a situation of greater stability. Attempts to get inert gases to form compounds, at the time they were discovered, failed, and chemists were quickly satisfied that this made sense.

To be sure, chemists continued to try, now and again, but they also continued to fail. Until 1962, then, the only successes chemists had had in tying the inert gas atoms to other atoms was in the formation of "clathrates." In a clathrate, the atoms making up a molecule form a cage-like structure and, sometimes, an extraneous atom—even an inert gas atom—is trapped within the cage as it forms. The inert gas is then tied to the substance and cannot be liberated without breaking down the molecule. However, the inert gas atom is only physically confined; it has not formed a chemical bond.

And yet, let's reason things out a bit. The boiling point of helium is 4.2° K,; that of neon is 27.2° K., that of argon 87.4° K., that of krypton 120.2° K., that of xenon 166.0° K. The boiling point of radon, the sixth and last inert gas and the one with the most massive atom, is 211.3° K. (−61.8° C.) Radon is not even a Gas, but merely a gas.

Furthermore, as the mass of the inert gas atoms increases, the ionization potential (a quantity which measures the ease with which an electron can be removed altogether from a particular atom) decreases. The increasing boiling point and decreasing ionization potential both indicate that the inert gases become less inert as the mass of the individual atoms rises.

By this reasoning, radon would be the least inert of the inert gases and efforts to form compounds should concentrate upon it as offering the best chance. However, radon is a radioactive element with a half-life of less than four days, and is so excessively rare that it can be worked with only under extremely specialized conditions. The next best bet, then, is xenon. This is very rare, but it is available and it is, at least, stable.

Then, if xenon *is* to form a chemical bond, with what other atom might it be expected to react? Naturally, the most logical bet would be to choose the most reactive substance of all—fluorine or some fluorine-containing compound. If xenon wouldn't react with that, it wouldn't react with anything.

(This may sound as though I am being terribly wise after the event, and I am. However, there are some who were legitimately wise. I am told that Linus Pauling reasoned thus in 1932, well before the event, and that a gentleman named A. von Antropoff did so in 1924.)

In 1962, Neil Bartlett and others at the University of British Columbia were working with a very unusual compound, platinum hexafluoride (PtF_6). To their surprise, they discovered that it was a particularly active compound. Naturally, they wanted to see what it could be made to do, and one of the thoughts that arose was that here might be something that could (just possibly) finally pin down an inert gas atom.

So Bartlett mixed the vapors of PtF_6 with xenon and, to his astonishment, obtained a compound which seemed to be $XePtF_6$, xenon platinum hexafluoride. The announcement of this result left a certain area of doubt, however. Platinum hexafluoride was a sufficiently complex compound to make it just barely possible that it had formed a clathrate and trapped the xenon.

A group of chemists at Argonne National Laboratory in Chicago therefore tried the straight xenon-plus-fluorine experiment, heating one part xenon with five parts of fluorine under pressure at 400° C. in a nickle container. They obtained xenon tetrafluoride (XeF_4), a straightforward compound of an inert gas, with no possibility of a clathrate. (To be sure, this experiment could have been tried years before, but it is no disgrace that it wasn't. Pure xenon is very hard to get and pure fluorine is very dangerous to handle, and no chemist could reasonably have been expected to undergo the expense and the risk for so slim-chanced a catch as an inert gas compound until after Bartlett's experiment had increased that "slim chance" tremendously.)

And once the Argonne results were announced, all Hades broke loose. It looked as though every inorganic chemist in the world went gibbering into the inert gas field. A whole raft of xenon compounds, including not only XeF_4, but also XeF_2, XeF_6, $XeOF_2$, $XeOF_3$, $XeOF_4$, XeO_3, H_4XeO_4, and H_4XeO_6 have been reported.

Enough radon was scraped together to form radon tetrafluoride (RnF_4). Even krypton, which is more inert than xenon, has been tamed, and krypton difluoride (KrF_2) and krypton tetrafluoride (KrF_4) have been formed.

The remaining three inert gases, argon, neon, and helium (in order of increasing inertness), as yet remain untouched. They are the last of the bachelors, but the world of chemistry has the sound of wedding bells ringing in its ears, and it is a bad time for bachelors.

As an old (and cautious) married man, I can only say to this—no comment.

16. THE HASTE-MAKERS

When I first began writing about science for the general public—far back in medieval times—I coined a neat phrase about the activity of a "light-fingered magical catalyst."

My editor stiffened as he came across that phrase, but not with admiration (as had been my modestly confident expectation). He turned on me severely and said, "Nothing in science is magical. It may be puzzling, mysterious, inexplicable—but it is *never magical.*"

It pained me, as you can well imagine, to have to learn a lesson from an *editor,* of all people, but the lesson seemed too good to miss and, with many a wry grimace, I learned it.

That left me, however, with the problem of describing the workings of a catalyst, without calling upon magical power for an explanation.

Thus, one of the first experiments conducted by any beginner in a high school chemistry laboratory is to prepare oxygen by heating potassium chlorate. If it were only potassium chlorate he were heating, oxygen would be evolved but slowly and only at comparatively high temperatures. So he is instructed to add some manganese dioxide first. When he heats the mixture, oxygen comes off rapidly at comparatively low temperatures.

What does the manganese dioxide do? It contributes no oxygen. At the conclusion of the reaction it is all still there, unchanged. Its mere presence seems sufficient to hasten the evolution of oxygen. It is a haste-maker or, more properly, a catalyst.

And how can one explain influence by mere presence? Is it a kind of molecular action at a distance, an extra-sensory perception on the part of potassium chlorate that the influential aura of manganese dioxide is present? Is it telekinesis, a para-natural action at a distance on the part of the manganese dioxide? Is it, in short, magic?

Well, let's see . . .

To begin at the beginning, as I almost invariably do, the first and most famous catalyst in scientific history never existed.

The alchemists of old sought methods for turning base metals into gold. They failed, and so it seemed to them that some essential ingredient was missing in their recipes. The more imaginative among them conceived of a substance which, if added to the mixture they were heating (or whatever) would bring about the production of gold. A small quantity would suffice to produce a great deal of gold and it could be recovered and used again, no doubt.

No one had ever seen this substance but it was described, for some reason, as a dry, earthy material. The ancient alchemists therefore called it *xerion,* from a Greek word meaning "dry."

In the eighth century the Arabs took over alchemy and called this gold-making catalyst "the xerion" or, in Arabic, *al-iksir*. When West Europeans finally learned Arabic alchemy in the thirteenth century, *al-iksir* became "elixir."

As a further tribute to its supposed dry, earthy properties, it was commonly called, in Europe, "the philosopher's stone." (Remember that as late as 1800, a "natural philosopher" was what we would now call a "scientist."

The amazing elixir was bound to have other marvelous properties as well, and the notion arose that it was a cure for all diseases and might very well confer immortality. Hence, alchemists began to speak of "the elixir of life."

For centuries, the philosopher's stone and/or the elixir of life was searched for but not found. Then, when finally a catalyst was found, it brought about the formation not of lovely, shiny gold, but messy, dangerous sulfuric acid.* Wouldn't you know?

Before 1740, sulfuric acid was hard to prepare. In theory, it was easy. You burn sulfur, combining it with oxygen to form sulfur dioxide. (SO_2). You burn sulfur dioxide further to make sulfur trioxide (SO_3). You dissolve sulfur trioxide in water to make sulfuric acid (H_2SO_4). The trick, though, was to make sulfur dioxide combine with oxygen. That could only be done slowly and with difficulty.

In the 1740s, however, an English sulfuric acid manufacturer named Joshua Ward must have reasoned that saltpeter (potassium nitrate), though nonflammable itself, caused carbon and sulfur to burn with great avidity. (In fact, carbon plus sulfur plus saltpeter is gunpowder.) Consequently, he added saltpeter to his burning sulfur and found that he now obtained sulfur trioxide without much trouble and could make sulfuric acid easily and cheaply.

The most wonderful thing about the process was that, at the end, the saltpeter was still present, unchanged. It could be used over and over again. Ward patented the process and the price of sulfuric acid dropped to 5 per cent of what it was before.

Magic?—Well, no.

In 1806, two French chemists, Charles Bernard Désormes and Nicholas Clément, advanced an explanation that contained a principle which is accepted to this day.

It seems, you see, that when sulfur and saltpeter burn together, sulfur dioxide combines with a portion of the saltpeter molecule to form a complex. The oxygen of the saltpeter portion of the complex transfers to the sulfur dioxide portion, which now breaks away as sulfur trioxide.

What's left (the saltpeter fragment minus oxygen) proceeds to pick up that missing oxygen, very readily, from the atmosphere. The saltpeter fragment, restored again, is ready to combine with an additional molecule of sulfur dioxide and pass along oxygen. It is the saltpeter's task sim-

* That's all right, though. Sulfuric acid may not be as costly as gold, but it is—conservatively speaking—a trillion times as intrinsically useful.

ply to pass oxygen from air to sulfur dioxide as fast as it can. It is a middleman, and *of course* it remains unchanged at the end of the reaction.

In fact, the wonder is not that a catalyst hastens a reaction while remaining apparently unchanged, but that anyone should suspect even for a moment that anything "magical" is involved. If we were to come across the same phenomenon in the more ordinary affairs of life, we would certainly not make that mistake of assuming magic.

For instance, consider a half-finished brick wall and, five feet from it, a heap of bricks and some mortar. If that were all, then you would expect no change in the situation between 9 A.M. and 5 P.M. except that the mortar would dry out.

Suppose, however, that at 9 A.M. you observed one factor in addition—a man, in overalls, standing quietly between the wall and the heap of bricks with his hands empty. You observed matters again at 5 P.M. and the same man is standing there, his hands still empty. He has not changed. However, the brisk wall is now completed and the heap of bricks is gone.

The man clearly fulfills the role of catalyst. A reaction has taken place as a result, apparently, of his mere presence and without any visible change of diminution in him.

Yet would we dream for a moment of saying "Magic!"? We would, instead, take it for granted that had we observed the man in detail all day, we would have caught him transferring the bricks from the heap to the wall one at a time. And what's not magic for the bricklayer is not magic for the saltpeter, either.

With the birth and progress of the nineteenth century, more examples of this sort of thing were discovered. In 1812, for instance, the Russian chemist Gottlieb Sigismund Kirchhoff . . .

And here I break off and begin a longish digression for no other reason than that I want to; relying, as I always do, on the infinite patience and good humor of the Gentle Readers.

It may strike you that in saying "the Russian chemist, Gottlieb Sigismund Kirchhoff" I have made a humorous error. Surely no one with a name like Gottlieb Sigismund Kirchhoff can be a Russian! It depends, however, on whether you mean a Russian in an ethnic or in a geographic sense.

To explain what I mean, let's go back to the beginning of the thirteenth century. At that time, the regions of Courland and Livonia, along the southeastern shores of the Baltic Sea (the modern Latvia and Estonia) were inhabited by virtually the last group of pagans in Europe. It was the time of the Crusades, and the Germans to the southeast felt it a pious duty to slaughter the poorly armed and disorganized pagans for the sake of their souls.

The crusading Germans were of the "Order of the Knights of the Sword" (better known by the shorter and more popular name of "Livonian Knights"). They were joined in 1237 by the Teutonic Knights, who had first established themselves in the Holy Land. By the end of the thirteenth century the Baltic shores had been conquered, with the German expeditionary forces in control.

The Teutonic Knights, as a political organization, did not maintain control for more than a couple of centuries. They were defeated by the Poles in the 1460s. The Swedes, under Gustavus Adolphus, took over in the 1620s, and in the 1720s the Russians, under Peter the Great, replaced the Swedes.

Nevertheless, however the political tides might shift and whatever flag flew and to whatever monarch the loyal inhabitants might drink toasts, the land itself continued to belong to the "Baltic barons" (or "Balts") who were the German-speaking descendants of the Teutonic Knights.

Peter the Great was an aggressive Westernizer who built a new capital, St. Petersburg* at the very edge of the Livonian area, and the Balts were a valued group of subjects indeed.

This remained true all through the eighteenth and nine-

* The city was named for his name-saint and not for himself. Whatever Tsar Peter was, a saint he was not.

teenth centuries when the Balts possessed an influence within the Russian Empire out of all proportion to their numbers. Their influence in Russian science was even more lopsided.

The trouble was that public education within Russia lagged far behind its status in western Europe. The Tsars saw no reason to encourage public education and make trouble for themselves. No doubt they felt instinctively that a corrupt and stupid government is only really safe with an uneducated populace.

This meant that even elite Russians who wanted a secular education had to go abroad, especially if they wanted a graduate education in science. Going abroad was not easy, either, for it meant learning a new language and new ways. What's more, the Russian Orthodox Church viewed all Westerners as heretics and little better than heathens. Contact with heathen ways (such as science) was at best dangerous and at worst damnation. Consequently, for a Russian to travel West for an education meant the overcoming of religious scruples as well.

The Balts, however, were German in culture and Lutheran in religion and had none of these inhibitions. They shared, with the Germans of Germany itself, in the heightening level of education—in particular, of scientific education—through the eighteenth and nineteenth centuries.

So it follows that among the great Russian scientists of the nineteenth century we not only have a man with a name like Gottlieb Sigismund Kirchhoff, but also others with names like Friedrich Konrad Beilstein, Karl Ernst von Baer, and Wilhelm Ostwald.

This is not to say that there weren't Russian scientists in this period with Russian names. Examples are Mikhail Vasilievich Lomonosov, Aleksandr Onufrievich Kovalevski, and Dmitri Ivanovich Mendeléev.

However, Russian officialdom actually preferred the Balts (who supported the Tsarist government under which they flourished) to the Russian intelligentsia itself (which frequently made trouble and had vague notions of reform).

In addition, the Germans were the nineteenth-century

scientists *par excellence,* and to speak Russian with a German accent probably lent distinction to a scientist. (And before you sneer at this point of view, just think of the American stereotype of a rocket scientist. He has a thick German accent, *nicht wahr?*—And this despite the fact that the first rocketman, and the one whose experiments started the Germans on the proper track [Robert Goddard], spoke with a New England twang.)

So it happened that the Imperial Academy of Sciences of the Russian Empire (the most prestigious scientific organization in the land) was divided into a "German party" and a "Russian party," with the former dominant.

In 1880 there was a vacancy in the chair of chemical technology at the Academy, and two names were proposed. The German party proposed Beilstein, and the Russian party proposed Mendeléev. There was no comparison really. Beilstein spent years of his life preparing an encyclopedia of the properties and methods of preparation of many thousands of organic compounds which, with numerous supplements and additions, is still a chemical bible. This is a colossal monument to his thorough, hard-working competence—but it is no more. Mendeléev, who worked out the periodic table of the elements, was, on the other hand, a chemist of the first magnitude—an undoubted genius in the field.

Nevertheless, government officials threw their weight behind Beilstein, who was elected by a vote of ten to nine.

It is no wonder, then, that in recent years, when the Russians have finally won a respected place in the scientific sun, they tend to overdo things a bit. They've got a great deal of humiliation to make up for.

That ends the digression, so I'll start over—

As the nineteenth century wore on, more examples of haste-making were discovered. In 1812, for instance, the Russian chemist Gottlieb Sigismund Kirchhoff found that if he boiled starch in water to which a small amount of sulfuric acid had been added, the starch broke down to a simple form of sugar, one that is now called glucose. This would not happen in the absence of acid. When it did happen in the presence of acid, that acid was not consumed but was still present at the end.

Then, in 1816, the English chemist Humphry Davy found that certain organic vapors, such as those of alcohol, combined with oxygen more easily in the presence of metals such as platinum. Hydrogen combined more easily with oxygen in the presence of platinum also.

Fun and games with platinum started at once. In 1823 a German chemist, Johann Wolfgang Döbereiner, set up a hydrogen generator which, on turning an appropriate stop-cock, would allow a jet of hydrogen to shoot out against a strip of platinum foil. The hydrogen promptly burst into flame and "Döbereiner's lamp" was therefore the first cigarette lighter. Unfortunately, impurities in the hydrogen gas quickly "poisoned" the expensive bit of platinum and rendered it useless.

In 1831 an English chemist, Peregrine Phillips, reasoned that if platinum could bring about the combination of hydrogen and of alcohol with oxygen, why should it not do the same for sulfur dioxide? Phillips found it would and patented the process. It was not for years afterward, however, that methods were discovered for delaying the poisoning of the metal, and it was only after that that a platinum catalyst could be profitably used in sulfuric acid manufacture to replace Ward's saltpeter.

In 1836 such phenomena were brought to the attention of the Swedish chemist Jöns Jakob Berzelius who, during the first half of the nineteenth century, was the uncrowned king of chemistry. It was he who suggested the words "catalyst" and "catalysis" from Greek words meaning "to break down" or "to decompose." Berzelius had in mind such examples of catalytic action as the decomposition of the large starch molecule into smaller sugar molecules by the action of acid.

But platinum introduced a new glamor to the concept of catalysis. For one thing, it was a rare and precious metal. For another, it enabled people to begin suspecting magic again.

Can platinum be expected to behave as a middleman as saltpeter does?

At first blush, the answer to that would seem to be in the negative. Of all substances, platinum is one of the most inert. It doesn't combine with oxygen or hydrogen under any normal circumstances. How, then, can it cause the two to combine?

If our metaphorical catalyst is a bricklayer, then platinum can only be a bricklayer tightly bound in a straitjacket.

Well, then, are we reduced to magic? To molecular action at a distance?

Chemists searched for something more prosaic. The suspicion grew during the nineteenth century that the inertness of platinum is, in one sense at least, an illusion. In the body of the metal, platinum atoms are attached to each other in all directions and are satisfied to remain so. In bulk, then, platinum will not react with oxygen or hydrogen (or most other chemicals, either).

On the surface of the platinum, however, atoms on the metal boundary and immediately adjacent to the air have no other platinum atoms, in the air-direction at least, to attach themselves to. Instead, then, they attach themselves to whatever atoms or molecules they find handy—oxygen atoms, for instance. This forms a thin film over the surface, a film one molecule thick. It is completely invisible, of course, and all we see is a smooth, shiny, platinum surface, which seems completely nonreactive and inert.

As parts of a surface film, oxygen and hydrogen react more readily than they do when making up bulk gas. Suppose, then, that when a water molecule is formed by the combination of hydrogen and oxygen on the platinum surface, it is held more weakly than an oxygen molecule would be. The moment an oxygen molecule struck that portion of the surface it would replace the water molecule in the film. Now there would be the chance for the formation of another water molecule, and so on.

The platinum does act as a middleman after all, through its formation of the monomolecular gaseous film.

Furthermore, it is also easy to see how a platinum catalyst can be poisoned. Suppose there are molecules to which the platinum atoms will cling even more tightly than to oxygen. Such molecules will replace oxygen wherever it is found on the film and will not themselves be replaced by any gas in the atmosphere. They are on the platinum surface to stay, and any catalytic action involving hydrogen or oxygen is killed.

Since it takes very little substance to form a layer merely one molecule thick over any reasonable stretch of surface, a catalyst can be quickly poisoned by impurities that are present in the working mixture of gases, even when those impurities are present only in trace amounts.

If this is all so, then anything which increases the amount of surface in a given weight of metal will also increase the catalytic efficiency. Thus, powdered platinum, with a great deal of surface, is a much more effective catalytic agent than the same weight of bulk platinum. It is perfectly fair, therefore, to speak of "surface catalysis."

But what is there about a surface film that hastens the process of, let us say, hydrogen-oxygen combination? We still want to remove the suspicion of magic.

To do so, it helps to recognize what catalysts *can't* do.

For instance, in the 1870's, the American physicist Josiah Willard Gibbs painstakingly worked out the application of the laws of thermodynamics to chemical reactions. He showed that there is a quantity called "free energy" which always decreases in any chemical reaction that is spontaneous—that is, that proceeds without any input of energy.

Thus, once hydrogen and oxygen start reacting, they keep on reacting for as long as neither gas is completely used up, and as a result of the reaction water is formed. We explain this by saying that the free energy of the water is less than the free energy of the hydrogen-oxygen mixture. The reaction of hydrogen and oxygen to form water is analogous to sliding down an "energy slope."

But if that is so, why don't hydrogen and oxygen molecules combine with each other as soon as they are mixed? Why do they linger for indefinite periods at the top of the energy slope after being mixed, and react and slide downward only after being heated?

Apparently, before hydrogen and oxygen molecules (each composed of a pair of atoms) can react, one or the other must be pulled apart into individual atoms. That requires an energy input. It represents an upward energy slope, before the downward slope can be entered. It is an "energy hump," so to speak. The amount of energy that must be put into a reacting system to get it over that energy hump is called the "energy of activation," and the concept was first advanced in 1889 by the Swedish chemist Svante August Arrhenius.

When hydrogen and oxygen molecules are colliding at ordinary temperature, only the tiniest fraction happen to possess enough energy of motion to break up on collision. That tiniest fraction, which does break up and does react, then liberates enough energy, as it slides down the energy slope, to break up additional molecules. However, so little energy is produced at any one time that it is radiated away before it can do any good. The net result is that hydrogen and oxygen mixed at room temperature do not react.

If the temperature is raised, molecules move more rapidly and a larger proportion of them possess the necessary energy to break up on collision. (More, in other words, can slide over the energy hump.) More and more energy is released, and there comes a particular temperature when more energy is released than can be radiated away. The temperature is therefore further raised, which produces more energy, which raises the temperature still further—and hydrogen and oxygen proceed to react with an explosion.

In 1894 the Russian chemist Wilhelm Ostwald pointed out that a catalyst could not alter the free energy relationships. It cannot really make a reaction go, that would not go without it—though it can make a reaction go rapidly that in its absence would proceed with only imperceptible speed.

In other words, hydrogen and oxygen combine in the absence of platinum but at an imperceptible rate, and the platinum haste-maker accelerates that combination. For water to decompose to hydrogen and oxygen at room temperature (without the input of energy in the form of an electric current, for instance) is impossible, for that would mean spontaneously moving up an energy slope. Neither platinum nor any other catalyst could make a chemical reaction move up an energy slope. If we found one that did so, then *that* would be magic.*

But *how* does platinum hasten the reaction it does hasten? What does it do to the molecules in the film?

Ostwald's suggestion (accepted ever since) is that catalysts hasten reactions by lowering the energy of activation of the reaction—flattening out the hump. At any given temperature, then, more molecules can cross over the hump and slide downward, and the rate of the reaction increases, sometimes enormously.

For instance, the two oxygen atoms in an oxygen molecule hold together with a certain, rather strong, attachment, and it is not easy to split them apart. Yet such splitting is necessary if a water molecule is to be formed.

When an oxygen atom is attached to a platinum atom and forms part of a surface film, however, the situation changes. Some of the bond-forming capabilities of the oxygen molecule are used up in forming the attachment to the platinum, and less is available for holding the two oxygen atoms together. The oxygen atom might be said to be "strained."

If a hydrogen atom happens to strike such an oxygen atom, strained in the film, it is more likely to knock it apart into individual oxygen atoms (and react with one of them) than would be the case if it collided with an oxygen atom free in the body of a gas. The fact that the oxygen molecule is strained means that it is easier to break apart, and that the energy of activation for the hydrogen-oxygen combination has been lowered.

Or we can try a metaphor again. Imagine a brick resting on the upper reaches of a cement incline. The brick should, ideally, slide down the incline. To do so, however, it must overcome frictional forces which hold it in place

* Or else we would have to modify the laws of thermodynamics.

against the pull of gravity. The frictional forces are here analogous to the forces holding the oxygen molecule together.

To overcome the frictional force one must give the brick an initial push (the energy of activation), and then it slides down.

Now, however, we will try a little "surface catalysis." We will coat the slide with wax. If we place the brick on top of such an incline, the merest touch will start it moving downward. It may move downward without any help from us at all.

In waxing the cement incline we haven't increased the force of gravity, or added energy to the system. We have merely decreased the frictional forces (that is, the energy hump), and bricks can be delivered down such a waxed incline much more easily and much more rapidly than down an unwaxed incline.

So you see that on inspection, the magical clouds of glory fade into the light of common day, and the wonderful word "catalyst" loses all its glamor. In fact, nothing is left to it but to serve as the foundation for virtually all of chemical industry and, in the form of enzymes, the foundation of all of life, too.

And, come to think of it, that ought to be glory enough for any reasonable catalyst.

17. THE SLOWLY MOVING FINGER

Alas, the evidences of morality are all about us; the other day our little parakeet died. As nearly as we could make out, it was a trifle over five years old, and we had always taken the best of care of it. We had fed it, watered it, kept its cage clean, allowed it to leave the cage and fly about the house, taught it a small but disreputable vocabulary, permitted it to ride about on our shoulders and eat at will from dishes at the table. In short, we encouraged it to think of itself as one of us humans.

But alas, its aging process remained that of a parakeet. During its last year, it slowly grew morose and sullen; mentioned its improper words but rarely; took to walking rather than flying. And finally it died. And, of course, a similar process is taking place within me.

This thought makes me petulant. Each year I break my own previous record and enter new high ground as far as age is concerned, and it is remarkably cold comfort to think that everyone else is doing exactly the same thing.

The fact of the matter is that I resent growing old. In my time I was a kind of mild infant prodigy—you know, the kind that teaches himself to read before he is five and enters college at fifteen and is writing for publication at eighteen and all like that there. As you might expect, I came in for frequent curious inspection as a sort of ludicrous freak, and I invariably interpreted this inspection as admiration and loved it.

But such behavior carries its own punishment, for the moving finger writes, as Edward Fitzgerald said Omar Khayyam said, and having writ, moves on. And what that means is that the bright, young, bouncy, effervescent infant prodigy becomes a flabby, paunchy, bleary, middle-aged non-prodigy, and age sits twice as heavily on such as these.

It happens quite often that some huge, hulking, raw-boned fellow, cheeks bristling with black stubble, comes to me and says in his bass voice, "I've been reading you ever since I learned to read; and I've collected all the stuff you wrote *before* I learned to read and I've read that, too." My impulse then is to hit him a stiff right cross to the side of the jaw, and I might do so if only I were quite sure he would respect my age and not hit back.

So I see nothing for it but to find a way of looking at the bright side, if any exists . . .

How long do organisms live anyway? We can only guess. Statistics on the subject have been carefully kept only in the last century or so, and then only for Homo sapiens, and then only in the more "advanced" parts of the world.

So most of what is said about longevity consists of quite rough estimates. But then, if everyone is guessing, I can guess, too; and as lightheartedly as the next person, you can bet.

In the first place, what do we mean by length of life? There are several ways of looking at this, and one is to consider the actual length of time (on the average) that actual organisms live under actual conditions. This is the "life expectancy."

One thing we can be certain of is that life expectancy is quite trifling for all kinds of creatures. If a codfish or an oyster produces millions or billions of eggs and only one or two happen to produce young that are still alive at the end of the first year, then the average life expectancy of all the coddish or oysterish youngsters can be measured in weeks, or possibly even days. I imagine that thousands upon thousands of them live no more than minutes.

Matters are not so extreme among birds and mammals where there is a certain amount of infant care, but I'll bet relatively few of the smaller ones live out a single year.

From the cold-blooded view of species survival, this is quite enough, however. Once a creature has reached sexual maturity, and contributed to the birth of a litter of young which it sees through to puberty or near-puberty, it has done its bit for species survival and can go its way. If it survives and produces additional litters, well and good, but it doesn't have to.

There is, obviously, considerable survival value in reaching sexual maturity as early as possible, so that there is time to produce the next generation before the first is gone. Meadow mice reach puberty in three weeks and can bear their first litter six weeks after birth. Even an animal as large as a horse or cow reaches the age of puberty after one year, and the largest whales reach puberty at two. Some large land animals can afford to be slower about it. Bears are adolescent only at six and elephants only at ten.

The large carnivores can expect to live a number of years, if only because they have relatively few enemies (always excepting man) and need not expect to be anyone's dinner. The largest herbivores, such as elephants and hippopotami, are also safe; while smaller ones such as baboons and water buffaloes achieve a certain safety by traveling in herds.

Early man falls into this category. He lived in small herds and he cared for his young. He had, at the very least, primitive clubs and eventually gained the use of fire. The average man, therefore, could look forward to a number of years of life. Even so, with undernourishment, disease, the hazards of the chase, and the cruelty of man to man, life was short by modern standards. Naturally, there was a limit to how short life could be. If men didn't live long enough, on the average, to replace themselves, the race would die out. However, I should guess that in a primitive society a life expectancy of 18 would be ample for species survival. And I rather suspect that the actual life expectancy of man in the Stone Age was not much greater.

As mankind developed agriculture and as he domesticated animals, he gained a more dependable food supply. As he learned to dwell within walled cities and to live under a rule of law, he gained greater security against human enemies from without and within. Naturally, life expectancy rose somewhat. In fact, it doubled.

However, throughout ancient and medieval times, I doubt that life expectancy ever reached 40. In medieval England, the life expectancy is estimated to have been 35, so that if you did reach the age of 40 you were a revered sage. What with early marriage and early childbirth, you were undoubtedly a grandfather, too.

This situation still existed into the twentieth century in some parts of the world. In India, for instance, as of 1950, the life expectancy was about 32; in Egypt, as of 1938, it was 36; in Mexico, as of 1940, it was 38.

The next great step was medical advance, which brought infection and disease under control. Consider the United States. In 1850, life expectancy for American white males was 38.3 (not too much different from the situation in medieval England or ancient Rome). By 1900, however, after Pasteur and Koch had done their work, it was up to 48.2; then 56.3 in 1920; 60.6 in 1930; 62.8 in 1940; 66.3 in 1950; 67.3 in 1959; and 67.8 in 1961.

All through, females had a bit the better of it (being the tougher sex). In 1850, they averaged two years longer life than males; and by 1961, the edge had risen to nearly seven years. Non-whites in the United States don't do quite as well—not for any inborn reason, I'm sure, but because they generally occupy a position lower on the economic scale. They run some seven years behind whites in life expectancy. (And if anyone wonders why Negroes are restless these days, there's seven years of life apiece that they have coming to them. That might do as a starter.)

Even if we restrict ourselves to whites, the United States does not hold the record in life expectancy. I rather think Norway and Sweden do. The latest figures I can find (the middle 1950s) give Scandinavian males a life expectancy of 71, and females one of 74.

This change in life expectancy has introduced certain changes in social custom. In past centuries, the old man was a rare phenomenon—an unusual repository of long memories and a sure guide to ancient traditions. Old age was revered, and in some societies where life expectancy is still low and old men still exceptional, old age is still revered.

It might also be feared. Until the nineteenth century there were particular hazards to childbirth, and few women survived the process very often (puerperal fever and all that). Old women were therefore even rarer than old men, and with their wrinkled cheeks and toothless gums were strange and frightening phenomena. The witch mania of early modern times may have been a last expression of that.

Nowadays, old men and women are very common and the extremes of both good and evil are spared them. Perhaps that's just as well.

One might suppose, what with the steady rise in life expectancy in the more advanced portions of the globe, that we need merely hold on another century to find men routinely living a century and a half. Unfortunately, this is not so. Unless there is a remarkable biological breakthrough in geriatrics, we have gone just about as far as we can go in raising the life expectancy.

I once read an allegory that has haunted me all my adult life. I can't repeat it word for word; I wish I could. But it goes something like this. Death is an archer and life is a bridge. Children begin to cross the bridge gaily, skipping along and growing older, while Death shoots at them. His aim is miserable at first, and only an occasional child is transfixed and falls off the bridge into the cloud-enshrouded mists below. But as the crowd moves farther along, Death's aim improves and the numbers thin. Finally, when Death aims at the aged who totter nearly to the end of the bridge, his aim is perfect and he never misses. And not one man ever gets across the bridge to see what lies on the other side.

This remains true despite all the advances in social structure and medical science throughout history. Death's aim has worsened through early and middle life, but those last perfectly aimed arrows are the arrows of old age, and even now they never miss. All we have done to wipe out war, famine, and disease has been to allow more people the chance of experiencing old age. When life expectancy was 35, perhaps one in a hundred reached old age; nowadays nearly half the population reaches it—but it is the same old old age. Death gets us all, and with every scrap of his ancient efficiency.

In short, putting life expectancy to one side, there is a "specific age" which is our most common time of death from inside, without any outside push at all; the age at which we would die even if we avoided accident, escaped disease, and took every care of ourselves.

Three thousand years ago, the psalmist testified as to the specific age of man (Ps. 90:10), saying: "The days of our years are threescore years and ten; and if by reason of strength they be fourscore years, yet is their strength labor and sorrow; for it is soon cut off, and we fly away."

And so it is today; three millennia of civilization and three centuries of science have not changed it. The commonest time of death by old age lies between 70 and 80.

But that is just the commonest time. We don't all die on our 75th birthday; some of us do better, and it is undoubtedly the hope of each one of us that we ourselves, personally, will be one of those who will do better. So what we have our eye on is not the specific age but the maximum age we can reach.

Every species of multicellular creature has a specific age and a maximum age; and of the species that have been studied to any degree at all, the maximum age would seem to be between 50 and 100 per cent longer than the specific age. Thus, the maximum age for man is considered to be about 115.

There have been reports of older men, to be sure. The most famous is the case of Thomas Parr ("Old Parr"), who was supposed to have been born in 1481 in England and to have died in 1635 at the age of 154. The claim is not believed to be authentic (some think it was a put-up job involving three generations of the Parr family), nor are any other claims of the sort. The Soviet Union reports numerous centenarians in the Caucasus, but all were born in a region and at a time when records were not kept. The old man's age rests only upon his own word, therefore, and ancients are notorious for a tendency to lengthen their years. Indeed, we can make it a rule, almost, that the poorer the recording of vital statistics in a particular region, the older the centenarians claim to be.

In 1948, an English woman named Isabella Shepheard died at the reported age of 115. She was the last survivor, within the British Isles, from the period before the compulsory registration of births, so one couldn't be certain to the year. Still, she could not have been younger by more than a couple of years. In 1814, a French Canadian named Pierre Joubert died and he, apparently, had reliable records to show that he was born in 1701, so that he died at 113.

Let's accept 115 as man's maximum age, then, and ask whether we have a good reason to complain about this. How does the figure stack up against maximum ages for other types of living organisms?

If we compare plants with animals, there is no question that plants bear off the palm of victory. Not all plants generally, to be sure. To quote the Bible again (Ps. 103:15-16), "As for man his days are as grass: as a flower of the field, so he flourisheth. For the wind passeth over it, and it is gone; and the place thereof shall know it no more."

This is a spine-tingling simile representing the evanescence of human life, but what if the psalmist had said that as for man his days are as the oak tree; or better still, as the giant sequoia? Specimens of the latter are believed to be over three thousand years old, and no maximum age is known for them.

However, I don't suppose any of us wants long life at the cost of being a tree. Trees live long, but they live slowly, passively, and in terribly, terribly dull fashion. Let's see what we can do with animals.

Very simple animals do surprisingly well and there are reports of sea-anemones, corals, and such-like creatures passing the half-century mark, and even some tales (not very reliable) of centenarians among them. Among more elaborate invertebrates, lobsters may reach an age of 50 and clams one of 30. But I think we can pass invertebrates, too. There is no reliable tale of a complex invertebrate living to be 100 and even if giant squids, let us say, did so, we don't want to be giant squids.

What about vertebrates? Here we have legends, particularly about fish. Some tell us that fish never grow old but live and grow forever, not dying till they are killed. Individual fish are reported with ages of several centuries. Unfortunately, none of this can be confirmed. The oldest age reported for a fish by a reputable observer is that of a lake sturgeon which is supposed to be well over a century old, going by a count of the rings on the spiny ray of its pectoral fin.

Among amphibia the record holder is the giant salamander, which may reach an age of 50. Reptiles are better. Snakes may reach an age of 30 and crocodiles may attain 60, but it is the turtles that hold the record for the animal kingdom. Even small turtles may reach the century mark, and at least one larger turtle is known, with reasonable certainty, to have lived 152 years. It may be that the large Galapagos turtles can attain an age of 200.

But then turtles live slowly and dully, too. Not as slowly as plants, but too slowly for us. In fact, there are only two classes of living creatures that live intensely and at peak level at all times, thanks to their warm blood, and these are the birds and the mammals. (Some mammals cheat a little and hibernate through the winter and probably extend their life span in that manner.) We might envy a tiger or an eagle if they lived a long, long time and even—as the shades of old age closed in—wish we could trade places with them. But do they live a long, long time?

Of the two classes, birds on the whole do rather better than mammals as far as maximum age is concerned. A pigeon can live as long as a lion and a herring gull as long as a hippopotamus. In fact, we have long-life legends about some birds, such as parrots and swans, which are supposed to pass the century mark with ease.

Any devotee of the Dr. Dolittle stories (weren't you?)

must remember Polynesia, the parrot, who was in her third century. Then there is Tennyson's poem *Tithonus,* about that mythical character who was granted immortality but, through an oversight, not freed from the incubus of old age so that he grew older and older and was finally, out of pity, turned into a grasshopper. Tennyson has him lament that death comes to all but him. He begins by pointing out that men and the plants of the field die, and his fourth line is an early climax, going, "And after many a summer dies the swan." In 1939, Aldous Huxley used the line as a title for a book that dealt with the striving for physical immortality.

However, as usual, these stories remain stories. The oldest confirmed age reached by a parrot is 73, and I imagine that swans do not do much better. An age of 115 has been reported for carrion crows and for some vultures, but this is with a pronounced question mark.

Mammals interest us most, naturally, since we are mammals, so let me list the maximum ages for some mammalian types. (I realize, of course, that the word "rat" or "deer" covers dozens of species, each with its own aging pattern, but I can't help that. Let's say the typical rat or the typical deer.)

Elephant	77	Cat	20
Whale	60	Pig	20
Hippopotamus	49	Dog	18
Donkey	46	Goat	17
Gorilla	45	Sheep	16
Horse	40	Kangaroo	16
Chimpanzee	39	Bat	15
Zebra	38	Rabbit	15
Lion	35	Squirrel	15
Bear	34	Fox	14
Cow	30	Guinea Pig	7
Monkey	29	Rat	4
Deer	25	Mouse	3
Seal	25	Shrew	2

The maximum age, be it remembered, is reached only by exceptional individuals. While an occasional rabbit may make 15, for instance, the average rabbit would die of old age before it was 10 and might have an actual life expectancy of only 2 or 3 years.

In general, among all groups of organisms sharing a common plan of structure the large ones live longer than the small. Among plants, the giant sequoia tree lives longer than the daisy. Among animals, the giant sturgeon lives longer than the herring, the giant salamander lives longer than the lizard, the vulture lives longer than the sparrow, and the elephant lives longer than the shrew.

Indeed, in mammals particularly, there seems to be a strong correlation between longevity and size. There are exceptions, to be sure—some startling ones. For instance, whales are extraordinarily short-lived for their size. the age of 60 I have given is quite exceptional. Most cetaceans are doing very well indeed if they reach 30. This may be because life in the water, with the continuous loss of heat and the never-ending necessity of swimming, shortens life.

But much more astonishing is the fact that man has a longer life than any other mammal—much longer than the elephant or even than the closely allied gorilla. When a human centenarian dies, of all the animals in the world alive on the day that he was born, the only ones that remain alive on the day of his death (as far as we know) are a few sluggish turtles, an occasional ancient vulture or sturgeon, and a number of other human centenarians. Not one non-human mammal that came into this world with him has remained. All, without exception (as far as we know), are dead.

If you think this is remarkable, wait! It is more remarkable than you suspect.

The smaller the mammal, the faster the rate of its metabolism; the more rapidly, so to speak, it lives. We might well suppose that while a small mammal doesn't live as long as a large one, it lives more rapidly and more intensely. In some subjective manner, the small mammal might be viewed as living just as long in terms of sensation as does the more sluggish large mammal. As concrete evidence of this difference in metabolism among mammals, consider the heartbeat rate. The following table lists some rough figures for the average number of heartbeats per minute in different types of mammals.

Shrew	1000	Sheep	75
Mouse	550	Man	72
Rat	430	Cow	60
Rabbit	150	Lion	45
Cat	130	Horse	38
Dog	95	Elephant	30
Pig	75	Whale	17

For the fourteen types of animals listed we have the heartbeat rate (approximate) and the maximum age (approximate), and by appropriate multiplications, we can determine the maximum age of each type of creature, not in years but in total heartbeats. The result follows:

Shrew	1,050,000,000
Mouse	950,000,000
Rat	900,000,000
Rabbit	1,150,000,000
Cat	1,350,000,000
Dog	900,000,000
Pig	800,000,000
Sheep	600,000,000
Lion	830,000,000
Horse	800,000,000
Cow	950,000,000
Elephant	1,200,000,000
Whale	630,000,000

Allowing for the approximate nature of all my figures, I look at this final table through squinting eyes from a distance and come to the following conclusion: A mammal can, at best, live for about a billion heartbeats and when those are done, it is done.

But you'll notice that I have left man out of the table. That's because I want to treat him separately. He lives at the proper speed for his size. His heartbeat rate is about that of other animals, of similar weight. It is faster than the heartbeat of larger animals, slower than the heartbeat of smaller animals. Yet his maximum age is 115 years, and that means his maximum number of heartbeats is about 4,350,000,000.

An occasional man can live for over 4 billion heartbeats! In fact, the life expectancy of the American male these days is 2.5 billion heartbeats. Any man who passes the quarter-century mark has gone beyond the billionth heartbeat mark and is still young, with the prime of life ahead.

Why? It is not just that we live longer than other mammals. Measured in heartbeats, we live *four times as long! Why??*

Upon what meat doth this, our species, feed, that we are grown so great? Not even our closest non-human relatives match us in this. If we assume the chimpanzee to have our heartbeat rate and the gorilla to have a slightly slower one, each lives for a maximum of about 1.5 billion heartbeats, which isn't very much out of line for mammals generally. How then do we make it to 4 billion?

What secret in our hearts makes those organs work so much better and last so much longer than any other mammalian heart in existence? Why does the moving finger write so slowly for us, and for us only?

Frankly, I don't know, but whatever the answer, I am comforted. If I were a member of any other mammalian species my heart would be stilled long years since, for it has gone well past its billionth beat. (Well, a *little* past.)

But since I am Homo sapiens, my wonderful heart beats even yet with all its old fire; and speeds up in proper fashion at all times when it should speed up, with a verve and efficiency that I find completely satisfying.

Why, when I stop to think of it, I am a young fellow, a child, an infant prodigy. I am a member of the most unusual species on earth, in longevity as well as brain power, and I laugh at birthdays.

(Let's see now. How many years to 115?)

Discus Books Distinguished Non-fiction

27979 $1.95

Dr. Asimov leads a breathtaking voyage of discovery through time and space that makes even the most complex theoretical concepts comprehensible, and that furnishes the basic knowledge with which to understand coming developments.

ILLUSTRATED WITH PHOTOGRAPHS AND DRAWINGS

On sale everywhere, or directly from the publisher. Include 25¢ for handling and mailing; allow 3 weeks for delivery.

AVON BOOKS, MAIL ORDER DEPT.
250 W. 55th St., N. Y., N. Y. 10019

UNIV 2-76